THE
PRINCIPLE OF INERTIA
IN THE MIDDLE AGES

ALLAN FRANKLIN 1938-

D1225658

COLORADO ASSOCIATED UNIVERSITY PRESS

COLORADO ASSOCIATED UNIVERSITY PRESS
1424 Fifteenth Street
Boulder, Colorado 80309

Library of Congress Catalog Card Number 76-10515
ISBN Number 0-87081-069-3

A condensed version of this paper originally appeared in The American
Journal of Physics, *Volume 44, Number 6, (June, 1976) pp. 529-545*
The author is grateful to Dr. Edwin F. Taylor, Editor of The Journal, *and*
to the American Association of Physics Teachers for permission
to reprint that material.

Cover Photograph: Wim Swaan.
Courtesy Paul Elek, Ltd.

To my parents, who would have been pleased.

CONTENTS

PREFACE

In dealing with the historical development of a physical principle, such as the principle of inertia, there is always the danger of giving too modern an interpretation of the work of previous writers. In general, I believe I have avoided this. As is shown in the text the medieval scientists did not have a clear or correct view of inertia. Their work, in many disparate threads, does, however, form an understandable bridge between the physics of Aristotle and that of Newton. I have examined what seemed to me to be the medieval precursors of the work of Galileo, Descartes, and Newton. In doing so I have been concerned primarily with the implications of the medieval works for future developments in physics rather than adhering strictly to the authors' own interpretations. In all probability, some of the medieval writers would be shocked at some of the conclusions drawn from their work. I believe, however, that the history presented is reasonable and plausible.

— —Allan Franklin
*Department of Physics and Astrophysics,
University of Colorado, Boulder*

INTRODUCTION

If some future historian of science were to try to reconstruct the history of mechanics from contemporary physics textbooks or even from some histories of physics he would surely arrive at a most unusual story. He would find an unexplained, and indeed unexplainable, gap of almost two thousand years between the work of Aristotle and that of Galileo. The revolution in mechanics in the seventeenth century would seem to occur out of thin air. The neglect of, and even hostility towards,[1] medieval science is perhaps understandable since, as Thomas Kuhn points out,[2] textbooks are always written after a revolution by victors who are notoriously ungenerous to losers. Although this revolution seems incomprehensible, it, like all other revolutions, had its roots in what came before.[3] Since twentieth-century historians of science[4] have made available to us the contributions of the medieval period this attitude on the part of modern textbook writers and, no doubt, physics teachers is not understandable.

We can only assume that their harsh comments about medieval science are based on biased, second-hand information since no one who has actually read the work of Buridan, Oresme, Bradwardine, Heytesbury, or Swineshead (to name only a few) can doubt either their ability as physicists or their contributions to physics.

Perhaps the most significant change in the transition from Aristotelian or medieval science to classical or Newtonian mechanics is seen in the contrasting phrasing of Aristotle's dictum *"Omne quod movetur ab alio movetur"* (all that is moved is moved by another) to Newton's First Law or Principle of Inertia that "Every body continues in its state of rest, or of uniform motion in a right line, unless it is compelled to change that state by forces impressed on it."[5] Thus, an object in motion, with no external forces acting on it such as resistance will continue forever. This is not a sudden change but rather one that has a long and detailed history in the medieval critiques of Aristotle's physics. The treatment of the problems of motion in a void, projectile motion, falling bodies, the inclined plane, and the cosmological question of the motion of the heavens and the earth by medieval scientists makes this transition quite understandable.

As we shall see, Aristotelian physics required a force in direct contact with the object for motion to occur at all. The resistance of a medium was also necessary to avoid motion that was infinite in both speed and extent, a consequence that was unacceptable to Aristotle. Obviously, then, motion in a void, where there was nothing to either move the object or to resist its motion, would be impossible.

Medieval critics of this view removed the resistance of a medium as a necessary condition for motion. Although they still require a force to produce and continue the motion, their idea of motion in a void without resistance is a first step toward the principle of inertia. The force required for an object moving in the void was provided by the original mover who impressed a force on the body. This impressed force was, however, self-expending to avoid motion that was infinite in extent, a conclusion the critics also rejected. This impressed force, both self-expending and later permanent, was used to explain not only motion in a void but also projectile motion and the motion of falling bodies. In particular, the idea of a permanent impressed force brings us quite close to the idea of permanent motion or the principle of inertia, although as we shall see the medieval views are not identical to it.

Aristotelian physics required a force to produce motion and regarded rest and motion as distinct states. In Newtonian physics a force causes a change in velocity, not velocity itself, and thus rest and uniform motion in a straight line are equivalent and indistinguishable. For this reason, the medieval discussions of the relativity of motion, particularly as they applied to the question of the motion of the heavens and the earth, are an important step towards an inertial mechanics.

We shall also deal with motion on an inclined plane. As Galileo noted, motion down such a plane is accelerated and motion up the plane is decelerated. Thus motion on a horizontal plane will be perpetual. Galileo's discussion is quite close to the principle of inertia but is, in fact, not quite correct. Galileo's work on inclined planes, as well as his other physics, owes much to his ancient and medieval predecessors.

ONE

ARISTOTELIAN PHYSICS

Medieval physics begins with the work of Aristotle (384-322 B.C.). It is hardly an exaggeration to say that medieval physics consists primarily of footnotes to Aristotle since virtually all of the important work in physics done in the Middle Ages is a commentary on Aristotelian physics.[6]

Aristotle categorizes local motion[7] as either natural motion or violent motion. He further divides natural motion into celestial motion, in which the natural motion is uniform and circular, and terrestrial motion in which the natural motion is rectilinear (either straight up or straight down). The terrestrial motions are governed by the nature of the four elements which make up all substances: earth, air, fire, and water, and the idea of natural place. Earth, which is absolutely heavy, falls toward the center of the universe (coincidentally the center of the earth) which is its natural place. Fire, which is absolutely light, rises to its natural place, a spherical shell inside the lunar sphere.[8] Objects made of mixtures of the elements behave in the manner of their predominant element. All motions which are not natural are classified as violent.

For both natural and violent motions Aristotle requires a force in direct contact with the object being moved. For living creatures the motive power is provided by the soul. Celestial objects and planets are moved by celestial intelligences or spirits. Inanimate objects require a force in contact with the object itself and, as we shall see below, this posed a difficult problem in dealing with both falling bodies and projectiles.

We may represent the Aristotelian law of motion by the modern formula:

$$\text{Velocity} = \frac{\text{Force (Motive power)}}{\text{Resistance}} \text{ or } \frac{V = k\,F}{R}$$

Aristotle never stated his law of motion in this concise form but rather discusses separately how either the distance travelled or the time of travel will change when either the force exerted on the object or the weight of the object changes.

> If then; A the movent have moved B a distance C in time D, then in the same time the force A will move ½ B twice the distance C, and in ½ D it will move ½ B the whole distance C: for thus the rules of proportion will be observed. Again if a given force move a given weight a certain distance in a certain time and half the distance in half the time, half the motive power will move half the weight the same distance in the same time.

Let E represent half the motive power A and Z half the weight B: then the ratio between the motive power and the weight in the one case is similar and proportionate to the ratio in the other, so that each force will cause the same distance to be travelled in the same time.[9]

Note here the non-inertial character of Aristotle's view, since velocity or motion requires a force to sustain it. We should also emphasize that Aristotelian motion required the action of *both* the force and the resistance as we shall see below.

Aristotle recognized that his law of motion did not have universal applicability. He knew that the application of a force might not result in any motion at all (even though his law implies this) and added the subsidiary condition that the force must be greater than the resistance.

If, then, (a force) A move B a distance C in time D, it does not follow that E, being half of A, will in the time D, or in any fraction of it cause B to traverse a part of C the ratio between which and whole of C is proportionate to that between A and E (whatever fraction of A E may be), in fact it might well be that it will produce no motion at all; otherwise one man might move a ship, since both the motive power of the ship-haulers and the distance that they all cause the ship to traverse are divisible into as many parts as there are men.[10]

He observed that one can exert a force and produce no motion at all, as in the case of someone attempting to lift a very large weight. This qualification seems to have been lost or misunderstood in the Middle Ages and forms the basis for some of the critiques of the Aristotelian law of motion.[11]

In applying his law of motion to falling bodies Aristotle associated the weight of the body with the force, and the resistance with the resistance of the medium. Thus Aristotle believed that heavy bodies fall faster than light ones. "We see that bodies which have a greater impulse either of weight or of lightness, if they are alike in other respects, move faster over an equal space, and in the ratio which their magnitudes bear to each other."[12]

The problem of what force it is that is actually in contact with the body and causes the body to fall posed a serious difficulty for Aristotle. It could not be what we would call the force of gravity, even though he associated the weight with the force, because this would involve action at a distance. Neither could it be an attraction to the body's natural place, as some later commentators claim, for the same reason. Aristotle was forced to conclude that whatever produced the body also gave it the tendency to fall to earth. The cause of the acceleration of a falling body was not dealt with by Aristotle but was considered by several of his medieval commentators.

Projectile motion posed a similar problem for Aristotle. In the case of a thrown rock the force was provided by the arm of the thrower as long as the body was in contact with it, but it was necessary to explain why the rock continued to move after it left the hand. Aristotle concluded that the medium provided the force to keep the projectile moving. This occurred either by "antiperistasis" (replacement) in which the medium rushes around to prevent the formation of a void and pushes the projectile from behind[13] or by the medium itself having acquired the power to be a mover from the original projector.

> Further, in point of fact, things that are thrown, move though that which gave them their impulse is not touching them, either by reason of mutual replacement, as some maintain, or because the air that has been pushed pushes them with a movement quicker than the natural locomotion of the projectile wherewith it moves to its proper place.[14]

Curiously the medium itself does not move but rather possesses the power to move something else. This power is, however, imperfectly transmitted from one layer of the medium to the next and gradually dies away. The resistance is provided by the resistance of the medium but also seems to include the mass or weight of the projectile. The resistance of the medium is needed to prevent the motion of the projectile from being infinite in both speed and extent, an impossibility for Aristotle. (See below.) As we shall see, several medieval critics of Aristotle noted his paradoxical use of the medium to both sustain and resist motion.

In view of the discussion above it is obvious that Aristotle regarded both the existence of a void or any motion in it as impossible. First, a void contains nothing which would be able to sustain the motion of a projectile once it has left the projector. Second, since a void can provide no resistance, the speed of the object would be infinite or as Aristotle says, "it moves through the void with a speed beyond any ratio."[15] He further argues that although heavy bodies fall faster than light ones in a medium, (see quote above) they would move with the same speed in a void.

> Therefore they will also move through the void with this ratio of speed. But this is impossible; for why should one move faster? (In moving through *plena* it must be so; for the greater divides them faster by its force. For a moving thing cleaves the medium either by its shape, or by the impulse which the body that is carried along or is projected possesses.) Therefore all will possess equal velocity. But this is impossible.[16]

From the point of view of a modern physicist and for the purposes of this paper, his most interesting and significant argument against motion in a void is as follows:

> Further, no one could say why a thing once set in motion should stop anywhere; for why should it stop *here* rather than *here*? So that a thing will either be at rest or must be moved *ad infinitum*, unless something more powerful get in its way.[17]

Aristotle has not only given a clear statement of the inertial principle but has also given a modern proof of it by relating the conservation of momentum to the homogeneity of space. He has, however, turned the argument around and regards it rather as a proof of the impossibility of motion in a void. The inertial consequences of this motion would be the unacceptable (to Aristotle) result of motion that is infinite in extent.

With this background in Aristotelian physics we can now examine the medieval precursors[18] of Newton and the inertial principle. We shall deal particularly with the medieval treatment of the problems of motion in a void, projectile motion, falling bodies, the relative motion of the heavens and the earth, and the inclined plane. We shall show that the medieval discussions of these problems lead in a very natural way to the inertial principle, although they are not themselves strictly inertial in character, but rather what we might call "pre-inertial." In particular we will demonstrate that these solutions had a direct influence on the work of Galileo at different stages in his career and influenced both his and subsequent views of inertia.

TWO

MOTION IN A VOID

Aristotle's views did not go unchallenged in the ancient world.[19] The atomists required the existence of a void, or unoccupied space, in which the atoms could move freely. Democritus, in fact, had stated that such a void must be infinite.[20] Lucretius added that all bodies, regardless of weight, would fall with the same speed in a void.

> For all things that fall through the water and thin air, these things must needs quicken their fall in proportion to their weights, just because the body of water and the thin nature of air cannot check each thing equally, but

---

[19] This section is based in part on two papers. Ernest A. Moody, "Galileo and Avempace: The Dynamics of the Leaning Tower Experiment," *Journal of the History of Ideas*, Vol. 12 (1951), pp. 163-93 and 375-422, (hereafter Moody) and Edward Grant, "Motion in the Void and the Principle of Inertia in the Middle Ages," *Isis*, Vol. 55 (1964), pp. 265-92, (hereafter Grant, *Motion in the Void*).

give place more quickly when overcome by heavier bodies. But, on the other hand, the empty void cannot on any side, at any time, support anything, but rather, as its own nature desires, it continues to give place; wherefore all things must needs be borne on through the calm void, moving at an equal rate with unequal weights.[21]

Aristotle's views on projectile motion were challenged by Hipparchus (second century B.C.) who attributed the motion to an impressed force which gradually diminishes.

Hipparchus, on the other hand, in his work entitled *On Bodies Carried Down by Their Weight* declares that in the case of earth thrown upward it is the projecting force that is the cause of the upward motion, so long as the projecting force overpowers the downward tendency of the projectile, and that to the extent that this projecting force predominates, the object moves more swiftly upward; then as the force is diminished (1) the upward motion proceeds but no longer at the same rate, (2) the body moves downward under the influence of its own internal impulse, even though the original force lingers in some measure, and (3) as this force continues to diminish the object moves downward more swiftly, and most swiftly when this force is entirely lost.[22]

Hipparchus then uses a similar theory to explain the acceleration of falling bodies released from rest. He claims that a residual force remains from whatever was restraining the body and this gradually dies away. This diminishing force, combined with the constant weight of the body, thus causes the acceleration. He is, of course, assuming here the Aristotelian view that velocity is proportional to force.

> Now Hipparchus asserts that the same cause operates in the case of bodies let fall from above. For, he says, the force which held them back remains with them up to a certain point, and this is the restraining factor which accounts for the slower movement at the start of the fall.[2 2]

The most important early medieval critic of Aristotle's view that the resistance of the medium was necessary for motion was the Christian neo-Platonist, John Philoponus (A.D. late fifth and early sixth century). He rejected the Aristotelian formula $V = kF/R$ and substituted $V = F - R$. Velocity is dependent not on the ratio of force to resistance but on their difference. Thus motion in a void, where the resistance is zero, becomes possible. In fact, since $V = F$ motion in a void measures the action of a force.

> For if a body moves the distance of a stade through air, and the body is not at the beginning and at the end of the stade at one and the same instant, a definite time will be

required, dependent on the particular nature of the body in question, for it to travel from the beginning of the course to the end (for, as I have indicated, the body is not at both extremities at the same instant), and this would be true even if the space traversed were in a void.[2 3] But a certain *additional time* is required because of the interference of the medium. For the pressure of the medium and the necessity of cutting through it make motion through it more difficult. . . .

If a stone move the distance of a stade through a void, there will necessarily be a time, let us say an hour, which the body will consume in moving the given distance. But if we suppose this distance of a stade filled with water, no longer will the motion be accomplished in one hour, but a certain additional time will be necessary because of the resistance of the medium. Suppose that for the division of water another hour is required, so that the same weight covers the distance through a void in one hour and through water in two. Now if you thin out the water, changing it into air, and if air is half as dense as water, the time which the body had consumed in dividing the water will be proportionately reduced. In the case of water the additional time was one hour. Therefore the body will move the same distance through air in an hour and a half. If, again, you make the air half as dense, the motion will be accomplished in an hour and a quarter. And if you continue indefinitely to rarefy the medium, you will decrease indefinitely the time required for the division of the medium, for example, the additional hour required in the case of water. But you will never completely eliminate this additional time, for time is indefinitely divisible.[24]

Philoponus opposes Aristotle's view that in a void all objects fall with equal speeds (one of his arguments against the existence of a void) and states that in a void velocity is proportional to weight.[25]

Weight, then, is the efficient cause of downward motion, as Aristotle himself asserts. This being so, given a distance to be traversed, I mean through a void where there is nothing to impede motion, and given that the efficient cause of the motion differs, the resultant motions will inevitably be at different speeds, even through a void.[26]

Philoponus also rejects the idea that the medium can both sustain and resist the motion of a projectile. His arguments rejecting the medium as the cause of projectile motion are indicative of the high quality of his thought. In rejecting "antiperistasis" he notes that the complicated motion required of the air is quite improbable.

For, on this, the air in question must perform

three distinct motions: it must be pushed forward by the arrow, then move back, and finally turn and proceed forward once more. Yet air is easily moved, and once set in motion travels a considerable distance. How, then, can the air, pushed by the arrow, fail to move in the direction of the impressed impulse, but instead, turning about, as by some command, retrace its course? Furthermore, how can this air, in so turning about, avoid being scattered into space, but impinge precisely on the notched end of the arrow, and again push the arrow on and adhere to it? Such a view is quite incredible and borders on the fantastic.[27]

He further argues that it is not the air that provides the motive power for a projectile. He notes that if this were so we would not need the stone to be in contact with the hand or the arrow with the bowstring.

For it would be possible, without such contact, to place the arrow at the top of a stick, as it were on a thin line, and to place the stone in a similar way, and then, with countless machines, to set a large quantity of air in motion behind these bodies. Now it is evident that the greater amount of air moved and the greater the force with which it is moved the more should this air push the arrow or stone, and further should it hurl them. But the fact is that even if you place the arrow or stone upon a line or point quite devoid of thickness and set in motion all the air behind the projectile with all possible force, the projectile will not be moved the distance of a single cubit.[28]

Having rejected the standard explanations of projectile motion, Philoponus explains it using an impressed force or borrowed power, similar to that of Hipparchus. This force will not only act in a void, but its effects will be greater than those in a medium.

Rather is it necessary to assume that some incorporeal motive force is imparted by the projector to the projectile, and that the air set in motion contributes either nothing at all or else very little to this motion of the projectile. If, then, forced motion is produced as I have suggested, it is quite evident that if one imparts motion "contrary to nature" or forced motion to an arrow or a stone the same degree of motion will be produced much more readily in a void than in a plenum. And there will be no need of any agency external to the projector. . . .[28]

This borrowed power will not persist indefinitely and will gradually die out even in a void. It will also be destroyed by the resistance of the medium and by the natural tendencies of the body. Philoponus' idea of motion in a void is tied to this concept of a self-expending borrowed power since otherwise the motion would be infinite——a conclusion he too rejected.

Philoponus' work seems to have been known to Arab scientists. In a tenth century commentary on Aristotle, Yahya ibn Adi refutes the idea that the air sustains projectile motion and offers instead the idea of impressed force.

> It is necessary to say that a power which is incorporeal comes from the projector into the projectile; it makes it (the projectile) arrive at the terminus [of its motion]; then it (the force) is dissipated.[29]

The notion of impressed force is continued in the work of Avicenna (A.D. 980 - 1037), best known as a physician and medical writer. In his "Book of the Healing of the Soul," a commentary on Aristotle, he writes:

> As for the case where there is [violent motion with the] separation of the moved [from the motor] like the projectile or that which is rolled, the scientists disagree in their opinions. There are some who hold that the cause lies in the tendency of the air which has been pushed to get behind the projectile and to unite there with a force which presses against that which is in front of it. There are others who say that the pusher pushes the air and the projectile together, but the air is more receptive to pushing and so it is pushed more swiftly and thus pulls that which has been

placed in it. And there are those who hold
that the cause is in that force which the
moved acquires from the mover and which
persists in it for a time until it is abolished by
the opposing force of that (medium) which
touches it and is displaced by it. And just as
the force is weakened in the projectile, so the
natural inclination (*mail*) and the action of
friction becomes dominant over it, and thus
the force is abolished and consequently the
projectile passes in the direction of its natural
inclination. . . . And some people have spoken
for the doctrine of "engendering." They say
that it is of the nature of movement that
[another] movement is engendered after
it; . . . But when we have verified the matter
we have found the most valid opinion to be
that of those who hold that the moved received an inclination (*mail*) from the mover.
The inclination is that which is perceived by
the senses to be resisting a forceful effort to
bring the natural motion to rest or to change
one violent motion into another.[30]

We note that the third opinion offered by Avicenna
corresponds closely to Philoponus' and Adi's view of
impressed force. Avicenna's argument against the air as
the cause of projectile motion is as follows, ". . . [But]
how is it possible that we can say that the air which has
returned behind the projectile has united and pressed
forward that which is in front of it? And what has
caused the movement which is forward in direction to

unite and push what is behind it?"[30] This is very similar to Philoponus' argument given above.

Avicenna, though, offers a different view in the *mail* theory. His *mail* resists changes in its state of motion making it similar to the *vis inertiae* of Newton.[31] Most importantly his theory differs from that of Philoponus in that his *mail* is seen as persisting indefinitely in a void.

> If the violent movement of the projectile is produced by a force operating in the void, it ought to persist, without annihilation or any kind of interruption.[32]

Avicenna states, therefore, that since such everlasting motions are not seen in nature, a void does not exist. As with Aristotle, we see that the inertial consequences of motion in a void are used as an argument against the existence of a void.[33]

Avicenna also attempts to give a quantitative treatment of *mail*. The action of the *mail* depends on the weight of the body on which it acts. Bodies moved by a given *mail* travel with velocities inversely proportional to their weights, and bodies moving with a given velocity in a medium will travel distances directly proportional to their weights.[34]

Abu'l-Barakat (died 1168), a follower of Avicenna, offered a theory of self-expending *mail* which is closer to the original view of Philoponus. He added, however, that the weakening of the violent *mail* combined with the increasing natural *mail* or gravity of the body produces the acceleration of a falling body.[35]

The productive cause of the violent *mail*, being resident in that which effects violent (projectile) movement, is separated from the moving body and so does not produce in it successive inclinations to replace the portion of *mail* weakened by resistance. On the contrary, the source of natural *mail* is found in the stone, (and thus) it supplies it with one *mail* after another. ... So long as the force (i.e. gravity) is acting outside the natural place of the body, it produces successive inclinations in such a way that the force of the *mail* increases throughout the duration of the movement.[36]

The theory of impressed force of Philoponus and the Arabic scientists as influential on subsequent developments in Europe is questionable. Philoponus' "Commentary on Aristotle's Physics" was not published until 1535 in a Greek text and in a Latin text not until 1542 and so cannot have had any influence on Scholastic science.[37] Philoponus' views were attacked by Simplicius, another sixth-century commentator, in "Digressions Against John the Grammarian" which he added to his own commentary on Aristotle's "Physics." No complete text of a medieval translation of Simplicius' commentary is known although a fragment does exist.[38] A translation of his commentary on "On the Heavens" was done in 1271 by William Moerbeke.[39] The portions of Avicenna's work which contain his theory of impressed force do not appear in any of the medieval translations of his writings. Alpetragius seems to have referred to

Avicenna's theory but the complete version was not translated until 1528, and the abbreviated translation by Michael Scot in 1217 seems too brief to have been of any use.[40] Nevertheless the theory of impressed force (although not Philoponus' self-expending kind) appears to have been known in the thirteenth century since both Roger Bacon and St. Thomas Aquinas offer refutations of it.[41]

Philoponus' view that motion in a void could exist and his rejection of the resistance of a medium as necessary for motion was known in the Middle Ages. This appears in the work of the Spanish Arab, Avempace (1106-1138).[42] His work was never translated into Latin but was known by reference to it in Averroes' (1126-1198) commentary on Aristotle's "Physics." Since virtually no one in the Middle Ages read Aristotle without a copy of Averroes' commentary close at hand, Avempace's views received wide circulation. Averroes reports his views as follows:

> Avempace, however, here raises a good question. For he says that it does not follow that the ratio of the motion of one and the same stone in water to its motion in air is as the ratio of the density of water to the density of air, unless we assume that the motion of the stone takes time only because it is moved in a medium. And if we make this assumption, it would imply that motion only takes time because of something resisting it——for the medium seems to impede the thing moved. And, if this were the case, then the heavenly

bodies would be moved instantaneously as they have no medium resisting them. And he says that the ratio of the rarity of water to the rarity of air is as the ratio of the retardation suffered by the moving body in water to the retardation suffered by it in air.

And these are his own words, in the Seventh Book of his work, where he says: "This resistance which occurs between the plenum and the body which is moved in it is that between which and the potency of the void Aristotle made a ratio in his Fourth Book. And what is believed because of his opinion is not so. For the ratio of water to air in density is not as the ratio of the motion of the stone in water to its motion in air. On the contrary, the ratio of the cohesive power of water to that of air is as the ratio of the retardation suffered by the moved thing due to the medium in which it is moved, namely water, to the retardation suffered by it when it is moved in air.

"For, if what certain thinkers have believed were true, then natural motion would be violent. Thus, if there were no resistance present, how could there be motion? For it would necessarily be instantaneous. Moreover, what then shall be said concerning circular motion [that is, the circular motion of the heavenly spheres]? No resistance is there, since there is absolutely no division there, and the place of a circle is always the same, so that it does not abandon one place and enter another. Therefore, it is necessary that circular motion occur instantaneously. Yet we observe in it the greatest slowness, as in the case of the motion of the fixed stars, and also the greatest speed, as in the case of the diurnal rotation. And this is only due to a difference in perfection between the mover and the thing moved. Therefore, when the mover is of greater perfection, that which is moved by it will be swifter; and when the mover is of less perfection, it will be nearer [in perfection] to the thing moved and the motion will be slower."

And these are his [Avempace's] words. And if that which he has said be conceded, then Aristotle's demonstration will be false. For, if the ratio of the rarity of one medium to the rarity of another medium is as the ratio of the retardation suffered by a moved thing in one of them to the retardation suffered by it in the other, and is not as the ratio of the motion itself, it will not follow that what is moved in a void is moved instantaneously. For if this [that is, Avempace's contention] were the case, then there would be subtracted from the mobile's motion only the retardation which affects it by reason of the medium, and its natural motion would remain. And as every motion involves time, that which is moved in a void is also necessarily moved in time and with a divisible motion. If this is so, nothing impossible follows. This, therefore is Avempace's question.[43]

Avempace seems to follow Philoponus' substitution of $V = F - R$ for Aristotle's $V = kF/R$,[44] although he does not include the idea of a self-expending impressed force. He also offers the example of celestial motion as an illustration of non-instantaneous motion without the resistance of a medium, an argument that appears often in Scholastic writers.

Before going on to discuss the further history of the ideas of Philoponus and Avempace I would like to examine the significance of these ideas in the history of the principle of inertia. Several historians of science, including Koyré[45] and Moody,[46] have argued that since these ideas share with Aristotle's the view that a force is necessary for motion, and that a uniform force produces uniform motion, they are in fact an explicit denial of the inertial principle, where uniform motion in a void implies the absence of a force. In fact, for Philoponus and Avempace, uniform motion in a void measures the action of a force. This argument is, of course, technically correct but misses a crucial point. It is certainly an important step toward the inertial principle to go from the Aristotelian view that the resistance of a medium is necessary for motion to the view that such resistance is not only unnecessary for motion but rather impedes it, and that motion in a void is at least hypothetically possible. As we shall see later, Descartes and Gassendi arrive at the correct statement of the principle of inertia from consideration of the idea of motion without resistance.

Avempace's views did not go unchallenged by either Averroes or later Scholastic writers. Averroes, in fact, refutes them immediately after their presentation. We shall not deal extensively with the critics of Avempace's views, although they do contribute substantially to some aspects of dynamics.[46] Their arguments against Avempace may be summarized briefly as 1) those which maintain the correctness of the Aristotelian law of motion $V = kF/R$ and the necessity of a medium for motion, and 2) those which criticize Avempace on metaphysical grounds for treating the "nature" of heavy bodies as something distinct from the matter of the bodies, thus describing the "nature" as that which is the mover, and the matter as that which is being moved. Averroes states that such a distinction is incorrect and that the self-motion of an inanimate body is impossible. He also argues that it is absurd to regard as "natural" that which never occurs (i.e., motion in a void), and thus make all observable motions which occur in a medium "violent."[47]

Avempace's views were, however, defended by several Scholastic writers. The earliest was St. Thomas Aquinas (ca. 1225-1274). Although he rejects the existence of a void on the same inertial grounds as Aristotle (that a body once set in motion in a void would continue indefinitely), he supports the arguments of Avempace on motion in a hypothetical void.

> But several difficulties arise against the opinion of Aristotle. The first of these is that it does not seem to follow that if a motion

occurs in a void it would bear no ratio in speed to a motion made in a plenum. Indeed, any motion has a definite velocity [arising] from a ratio of motive power to mobile—— even if there should be no resistance. This is obvious by example and reason. An example is that of the celestial bodies, whose motions are not impeded by anything, and yet they have a definite speed in a definite time. An appeal to reason is this: Just as there is a prior and posterior part in a magnitude traversed by a motion so also we understand that in the motion (itself) there is a prior and posterior. From this it follows that motion takes place in a definite time. But it is true that in virtue of some impediment (or resisting medium) something could be subtracted from this speed. It is not necessary, therefore, that a ratio of speeds be related as a ratio of resis- tance to resistance, for then, if there were no resistance, motion would occur instantane- ously. But it is necessary that the ratio of retardation to retardation be as the ratio of resistant medium to resistant medium. Thus if motion in a void were assumed, it follows that n o retardation would occur beyond the natural velocity, and it does not follow that motion in a void would bear no ratio to motion in a plenum.[48]

Aquinas' main argument is that motion in a void would still take time since the body would still have to travel a distance. This argument was used by many adherents of motion in a void and was referred to as the *distantia terminorum* or *incompossibilitas terminorum*. This is quite similar to the argument given by Philoponus (see above). Aquinas also follows Avempace's idea of subtracting the resistance rather than dividing by it and uses the same example of celestial bodies as finite motion without resistance. Aquinas felt, however, that even motion in a void required the action of both a motive power and a resistance. The resistance is provided by the *corpus quantum* or magnitude or dimension of the body. As Grant points out,[49] once this *corpus quantum* is set in motion in a void, it would continue forever. Thus, although Aquinas rejects the existence of a void because of its inertial consequences, the same consequences seem to be implied in his discussion of motion in a hypothetical void.

The study of motion in a void received further encouragement in the Edict of Paris issued by Bishop Etienne Tempier in 1277. This condemnation of 219 separate theses was, in part, a conservative clerical reaction to certain scholars who maintained that God was restricted to creating the world in accordance with the principles of Aristotelian physics. Of particular interest is article 49 where the following thesis is condemned: "That God could not move the heavens with rectilinear motion; and the reason is that a vacuum would remain."[50] The entire condemnation is significant since it offers a theologically certified criticism of the infallibility of Aristotelian physics and thus encourages physical speculation. We have seen, however, that important criticism of Aristotle existed well before the

issuance of this edict. Article 49 is of particular importance because it changes the idea of a void from merely
a hypothetical to a supernatural possibility. In addition
by allowing the possibility of rectilinear motion of the
heavens it is a step toward the breakdown of the hierarchical distinction between celestial and terrestrial
motion, a point we shall return to later. Although the
penalty of excommunication for holding any of the
theses was lifted in 1325, the Church maintained its
neutrality on the issues, so that several fourteenth-
century writers refer to the articles.

The discussion of motion in a void is continued by
Peter John Olivi (died 1298), who criticizes Aristotle as
follows:

> In disproof of the second argument it should
> be known, first of all, that the argument of
> Aristotle is not valid——although those people
> who enslave their minds to him as if to a god
> believe this and any other argument of his,
> however sophistical, to be most perfect——just
> because it has been asserted and written down
> by their god.
>
> That, however, it is not valid, is evident in
> three ways. First, because when several causes
> are given to the same effect, of which one is
> more primary and fundamental and sufficient
> and necessary, then the elimination of the
> others, so long as this one remains, does not
> destroy that effect totally, but only in the
> respect of the increment which was received
> from the other causes. Thus it is, however, in

this case: because the basic cause of succession in local motion is the extended character of the space over which the motion takes place, and the divisible or extended quantity of the mobile body, and its inability to be simultaneously in several places;

On the other hand, the resistance of the mobile body, or of the bodies existing in the space through which it is moved, are secondary causes of the aforementioned succession, such that it is made slower and more retarded—*just as the deficiency of the motive power . . . is also a cause of greater slowness.* (Italics added) Therefore, if the principal cause remains, in the absence of these three, there always remains the necessity of a successive character in the passage of the mobile body over that space. . . .[51]

Olivi attacks not only Aristotle's physics but also his infallibility. We have seen that this is not unusual in this period although the vehemence of the attack is somewhat unusual. Olivi follows Aquinas in giving the *distantia terminorum* as the main argument in favor of motion in a void and in stating the secondary effect of the resistance of the medium. He goes further, in the italicized passage, and seems to regard the motive power as a secondary cause of motion, and thus unnecessary. This would seem to be at least an indication of the principle of inertia since motion could thus occur without a force. It is clear, however, that Olivi did not really mean this. In his discussion of projectile motion he requires a "violent impulse or inclination" to explain the continuance of the motion: " . . . and in the same way the motion of projectiles follow necessarily on the violent impulses or inclinations given to them by the thrower."[5][2]

An interesting continuation of the idea of motion in a void occurs in Duns Scotus (1266-1308) in his discussion of whether angels can undergo local motion. He argues that even an angel cannot be present in more than one part of space at a time and thus for the successive nature of motion.[5][3]

We have seen that some of the ideas of Avempace and Philoponus were well known to Scholastic writers. Not much progress was made, however, from Aquinas to Duns Scotus. In particular, Philoponus' idea of a self-expending impressed force does not appear in any of this work. Roger Bacon and Aquinas do, however, reject the idea of an impressed force to explain projectile motion, but it is not of the self-expending variety. Neither is Olivi's "violent impulse or inclination." The concept of a self-expending impressed force does surface again in the work of two followers of Duns Scotus, Franciscus de Marchia and Nicholas Bonetus.[5][4]

Franciscus de Marchia (around 1320) arrived at his discussion of projectile motion from the consideration of a theological problem, that of instrumental causality. He discusses whether grace comes only directly from God or whether some supernatural power to give grace remains in the Sacraments. In supporting the latter view he uses the motion of a projectile after it has left the projector as an analogy. He first examines, "whether in a stone thrown upward . . . there is received some force which continues the motion."[55] After rejecting the standard explanations he concludes " . . . motion of this kind arises immediately from some force left behind (*virtus derelicta*) by means of an initial action of the first motor, for example, a hand."[56] He then asks whether this force resides in the body itself or, as Aristotle says, in the medium.

> Whence it is to be known that the force moving some heavy body upward is twofold: one which begins the motion . . . and this force is the force of the hand; another force which comes after the motion has begun and continues it——and this is caused or left behind by the first [force] with the object of producing motion. For unless some force other than the first one is posited, it is impossible to give a cause for the succeeding motion, as it was deduced above. . . . And if one asks what sort is a force of this kind, it can be answered that it is not simply permanent nor simply fluent, but almost medial, since it lasts for a certain time——just as calidity generated by a fire does not have to be simply permanent as fire does. . . . It seems preferable that a force of this kind resides in the body which is moved rather than in the medium, regardless of what the Philosopher (Aristotle) and the Commentator (Averroes) have said on this matter. [This theory is preferred] because it would be in vain that something should be done by many [causes]

which can be done by few . . . (and) because
in positing this all phenomena are accounted
for better and more easily.[57]

Thus de Marchia has revived the theory of a self-
expending force to explain projectile motion. He does,
however, allow a secondary role for some force left in
the medium. He further states, "From this it follows
that with the intelligences ceasing to move the heavens,
the heavens would still be moved, or revolve for a time
by means of a force of this kind following and continuing the motion, as is evident in a potter's wheel
which revolves for a time after the first mover has
ceased to move [it]."[58] We see again, as in the Edict of
Paris, a beginning of the breakdown of the distinction
between celestial and terrestrial motions since de
Marchia is describing the motion of the heavens and of a
potter's wheel with the same physics.

Nicholas Bonetus (died 1343) extended the use of
this self-expending force to deal with motion in a void.
Bonetus argues, as others, that motion in a void would
be successive (*distantia terminorum*) and that the resistance of a medium just reduces the natural velocity. He
is left with the problem of explaining what would move
a projectile in the void. He answers that "some hold that
a violent motion can occur in a void without a real or
virtual conjunction of a prime moving or projecting
force with a mobile. The reason given for this is that in a
violent motion some non-permanent and transient form
is impressed on the mobile so that the motion in a void
is possible as long as this form endures, but when it
disappears the motion ceases."[59] Once again the inertial

consequences of motion in a void are avoided.

As argued earlier, although the views of Avempace, Philoponus, and their adherents are not inertial, since a force is still required for motion, they are an important first step. They remove the necessity of both the medium and its resistance for motion and consider the possibility of motion in a void, without resistance. That they are not inertial is further illustrated by the concept of a self-expending rather than a permanent force which explains projectile motion in both a plenum and a void. Avicenna did arrive at a concept of a permanent force in a void and we shall see a reappearance of this view when we consider the work in the fourteenth century. In addition, we have seen celestial motions given as an example of finite motion without resistance, and a beginning of the breakdown of the hierarchical distinction between celestial and terrestrial motions which would allow discussion of perpetual motion on earth. We shall return to both of these discussions later.

Before considering the criticism and modification of these views in the fourteenth century it is worth looking at the influence of these particular ideas in the early work of Galileo (1564-1642). There can be no doubt that Galileo was influenced by the tradition of Philoponus and Avempace. Galileo seems to acknowledge this in his discussion of motion in a void when he states, "And though Scotus, Saint Thomas, Philoponus, and some others hold a view opposed to Aristotle's they arrive at the truth by belief rather than by real proof or by refuting Aristotle."[60] This is patently false since, as we shall see, Galileo himself uses arguments virtually identical to those of his predecessors.[61] To give Galileo the benefit of the doubt, he may not have been aware of the complete content of Philoponus' work although it is dealt with in the writings of his teacher at Pisa, Francisco Bonamico. It is true, however, that several editions of Philoponus' work had been published in the sixteenth century. Moody[62] has given a detailed discussion of how Galileo may have become aware of the previous work. Regardless of how Galileo did learn of these ideas, there is no doubt of their influence on his Pisan dynamics. It is surprising, however, that Galileo does not seem to be aware, at this time, of the important criticism of Avempace's work and the further developments in mechanics that took place in the fourteenth century. We have argued previously that the views of Philoponus and Avempace are a significant step toward the inertial principle. Thus, we see the beginnings of Galileo's inertial views, which will, of course, be modified in his later work. He does, however, always retain his view that the medium is unnecessary for motion.

Galileo's earliest work "De Motu"[63] evidences clearly his indebtedness to earlier scientists. He follows Avempace and Philoponus in substituting $V = F - R$ for Aristotle's $V = kF/R$, although he uses the specific weights or densities of the body and the medium for the force and the resistance, in considering falling bodies.[64] He also seems indebted to Archimedes for both his content and his axiomatic methodology. His first reference to Philoponus' law of motion occurs in considering the upward motion of objects in water, a dynamic application of Archimedes' Principle.

If, for example, a piece of wood whose weight
is 4 moves upward in water, and the weight of
a volume of water equal to that of the wood
is 6, the wood will move with a speed we may
represent as 2. But if, now, the same piece of
wood is carried upward in a medium heavier
than water, a medium such that a volume of it
equal to the volume of wood has a weight of
10, the wood will rise in this medium with a
speed that we may represent as 6. But it is
moved in the other medium with a speed 2.
Therefore the two speeds will be to each
other as 6 and 2, and not (as Aristotle held) as
the weights or densities of the media which
are to each other as 10 and 6. It is clear, then,
that in all cases the speeds of upward motion
are to each other as the excess of weight of
one medium over the weight of the moving
body is to the excess of weight of the other
medium over the weight of the body.[65]

The same analysis is then applied to falling bodies using
a similar numerical example.

For, clearly, in the case of the same body
falling in different media, the ratio of the
speeds of the motions is the same as the ratio
of the amounts by which the weight of the
body exceeds the weights [of an equal
volume] of the respective media.[66]

He is now in a position to argue that motion in a void

would not be instantaneous, just as Philoponus and Avempace had done.

> Therefore, if, as Aristotle held, the ratio of the speeds were equal to the ratio, in the geometric sense, of the rarenesses of the media, Aristotle's conclusion would have been valid, that motion in a void could not take place in time. For the ratio of the time in the plenum to the time in the void cannot be equal to the ratio of the rareness of the plenum to the rareness of the void, since the rareness of the void does not exist. But if the ratio of the speeds were made to depend on the aforesaid ratio, not in the geometric, but in the arithmetic sense [i.e., as a ratio of the differences], no absurd conclusion would follow. And, in fact, the ratio of the speeds does depend, in an arithmetic sense, on the relation of the lightness of the first medium to that of the second. For the ratio of the speeds is equal, not to the ratio of the light-ness of the first medium to that of the second, but, as has been proved, to the ratio of the excess of the weight of the body over the weight of the first medium to the excess of the weight of the body over the weight of the second medium.[6 7]

Once he has established that motion in a void is possible he must then explain how projectiles are moved both in a medium and a void. He rejects the Aristotelian view that the medium sustains the motion by arguments similar to those of Philoponus and others we have discussed. He then states his own view that the motion is due to the action of a self-expending impressed force, using views similar to those of Hipparchus, Philoponus, and Franciscus de Marchia.

> But now, in order to explain our own view, let us first ask what is that motive force which is impressed by the projector upon the projectile. Our answer, then, is that it is a taking away of heaviness when the body is hurled upward, and a taking away of lightness, when the body is hurled downward. But if a person is not surprised that fire can deprive iron of cold by introducing heat, he will not be surprised that the projector can, by hurling a heavy body upward, deprive it of heaviness and render it light.
>
> The body, then, is moved upward by the projector so long as it is in his hand and is deprived of its weight; in the same way the iron is moved, in an alternative motion, towards heat, so long as the iron is in the fire and is deprived by it of coldness. Motive force, that is to say lightness, is preserved in the stone, when the mover is no longer in contact; heat is preserved in the iron after the iron is removed from the fire. The impressed force gradually diminishes in the projectile when it is no longer in contact with the projector; the heat diminishes in the iron, when the fire is not present.[68]

THREE

THE FOURTEENTH CENTURY

Most late thirteenth- and early fourteenth-century commentators sided with Averroes and Aristotle against Avempace and Aquinas without adding too much of significance to the discussion. The first important criticism of both views was done by William of Ockham (1300-1350).[69] Ockham seems to have been the first writer to separate the problem of kinematics from that of dynamics. By kinematics we mean the definition and measurement of motion. Dynamics is the measurement of forces and their effects by the work done. It is also the discussion of forces as the causes of motion while kinematics is the description of motion. Ockham's discussions seem to contain parts of the views of Averroes and Aristotle as well as parts of the ideas of Avempace and Aquinas. He agrees with Aquinas that motion in a void, without resistance, would be successive because of the distance travelled (*distantia terminorum*). He sides with Averroes in his notion that where resistance is present the time of motion depends on the ratio of the motive power to the resistance.

It is to be said that the Commentator does not argue against Avempace because of his assumption that distance between initial and terminal positions (*distantia terminorum*) would of itself suffice to determine that rectilinear motion should take time; but he argues against him for supposing that each heavy and light body has one natural motion, to which there is added, as if to something distinct, an accidental retardation due to the resistance of the medium——in the manner in which one line is added to another. And against this view the argument of the Commentator holds good; because then (i.e., in Avempace's theory) the proportion of time to time would not be according to the proportion of rarity of the medium to rarity nor according to the proportion of resistance to resistance.

And thus it is explicitly evident that the Commentator refutes Avempace because of his assumption that the retardation from the resistance of the medium is something distinct added to the natural motion; and this I do not assume, and hence his argument is not against me because of my further assumption that a mobile body could be moved in time, even though the medium did not resist in a positive sense. But it is compatible with this, that when there is a resistant medium then the proportion of resistance to resistance is as that of time to time and of motion to motion, and conversely. Consequently the Commentator does not argue against Avempace's assumption that motion can take place in time solely by reason of spatial distance.[70]

Most importantly, Ockham goes on to attack the implicit view of both Avempace and Averroes that "everything that is moved is moved by another." For Ockham, only *res absolutae* or *res permanentes*, individual things, determined by observable qualities exist. "Apart from *res absolutae*, that is substances and qualities, no thing is imaginable either in actuality or potentiality."[71] Motion, which he defined as an object having successive existence, without intermediate rest, in different places is not a reality apart from the moving bodies.

Motion is not such a thing wholly distinct in itself from the permanent body, because it is futile to use more entities when it is possible to use fewer. . . . That without such an additional thing we can save motion and everything that is said about motion, is made clear by considering the separate parts of motion. For it is clear that local motion is to be conceived as follows: positing that the body is in one place and later in another place, thus proceeding without any rest or any intermediate thing, other than the body itself and the agent itself which moves we have local motion truly. Therefore it is futile to postulate such other things.[72]

We shall see later that Descartes held similar views on motion. Since motion is not a real or a new effect for Ockham it does not require a cause either from the medium or from an impressed force.

> I say therefore, that that which moves in motion of this kind, after the separation of the moving body from the projector, is the body moved by itself and not by any power in it or relative to it, for it is impossible to distinguish between that which does the moving and that which is moved. If you say that a new effect has some cause and that local motion is a new effect, I say that local motion is not a new effect in the sense of a real effect . . . because it is nothing else but the fact that the moving body is in different parts of space in such a manner that it is not in any one part, since two contradictories cannot both be true. . . . It would indeed be astonishing if my hand were to cause some power in the stone by the mere fact that through local motion it came into contact with the stone.[73]

Ockham's views are yet another important step toward the principle of inertia. Obviously, if motion does not require a cause, once it exists it may continue forever. Crombie[74] has pointed out that Ockham's views are not, however, equivalent to the inertial principle. For both Descartes and Newton the change from a state of rest to a state of motion is a new effect requiring a cause, namely a force. Force may not be necessary to produce velocity but it is required to change velocity as, for example, going from rest to motion.

Ockham's discussion is continued by Marsilius of Inghen (ca. 1340-1396) in his "Questions on the Eight Books of the Physics [in the Nominalist Manner]." Marsilius agrees with the view that local motion is to be identified with the body being moved. In addition, he discusses the idea of an infinite void space.

> Therefore, they posit that there is place outside of the heavens, or an infinite space. Therefore if God was to move the whole world rectilinearly or circularly, the world would be differently disposed with respect to the place or separate space in which it would be.[75]

We have seen earlier that the lack of a concept of an infinite space led several writers, notably Aristotle and Philoponus, to reject the inertial consequences of motion in a void. In a finite space motion cannot continue forever. Therefore discussions of an infinite space may also be considered a step toward inertia. We see also the influence of Article 49 of the Edict of Paris. A similar influence appears in other fourteenth-century writers on the void, including Bradwardine (1290-1349) and Oresme (1325-1382).

In his "De causa Dei contra Pelagium" Bradwardine discusses the properties of God as follows:

1. First, that, essentially and in presence,

God is necessarily everywhere in the world and all its parts.

2. And also beyond the real world in a place or imaginary infinite void.

3. And it also seems obvious that a void can exist without body, but in no manner can it exist without God.[76]

Oresme states a similar view in his discussion of "whether there may be or could be something outside the heaven." He responds "But, according to faith there is no space outside the heaven but we can concede that outside the heaven there may be a vacuum because God can create a body or a place there. Therefore, if it is asked what is that vacuum outside the heaven, one should reply that it is nothing but God Himself, Who is His own indivisible immensity and His own eternity as a whole and all at one."[77] These discussions had an influence on the seventeenth-century considerations of the void[78] and should also be considered steps toward inertia.

As we have seen earlier, William of Ockham[79] rejected the idea of an impressed force to explain projectile motion. This rejection was shared in the period 1290 to 1320 in the work of Aegidius of Rome, Richard of Middleton, John of Jandun, and others, who argued that the medium must supply the power for the projectile. The notion of impressed force was also rejected by all of the Merton College writers (see Ref. 79). In the early 1320's the theory of self-expending impressed force had been revived by Franciscus de Marchia. This work may have influenced the work of John Buridan (1300-1358). Buridan also uses a theory of impressed force to explain projectile motion. Buridan rejects air as the motive power for projectiles and states,

> Thus we can and ought to say that in the stone or other projectile there is impressed something which is the motive force (*virtus motiva*) of that projectile. And this is evidently better than falling back on the statement that the air continues to move that projectile. For the air appears rather to resist. Therefore, it seems to me that it ought to be said that the motor in moving a moving body impresses in it a certain impetus (*impetus*) or a certain motive force (*vis motiva*) of the moving body, (which impetus acts) in the direction toward which the mover was moving the moving body, either up or down, or laterally or circularly.[80]

Buridan's impetus is quite different from Marchia's impressed force. The impressed force lasts only for a certain time, while Buridan regards impetus as permanent unless acted on by resistance or other forces.

> The third conclusion is that that impetus is a thing of permanent nature (*res nature permanentis*), distinct from the local motion in which the projectile is moved. . . . And it is probable that that impetus is a quality

naturally present and predisposed for moving a body in which it is impressed, just as it is said that a quality impressed in iron by a magnet moves the iron to the magnet. And it is also probable that just as that quality (the impetus) is impressed in the moving body along with the motion by the motor; so with the motion it is remitted, corrupted, or impeded by resistance or a contrary inclination.[8][1]

Impetus is also given a quantitative definition by Buridan. It is proportional to both the quantity of matter (or mass) in the object and to its speed.

And by the amount the motor moves that moving body more swiftly, by the same amount will it impress in it a stronger impetus .... Hence by the amount more there is of matter, by that amount can the body receive more of that impetus and more intensely. Now in a dense and heavy body, other things being equal, there is more of prime matter than in a rare and light one. Hence a dense and heavy body receives more of that impetus and more intensely, just as iron can receive more calidity than wood or water of the same quantity. . . . And so also if light wood and heavy iron of the same volume and of the same shape are moved equally fast by a projector, the iron will be moved farther because there is impressed in it a more intense impetus, which is not so quickly corrupted as the lesser impetus would be corrupted. This is also the reason why it is more difficult to bring to rest a large smith's mill which is moving swiftly than a small one, evidently because in the large one, other things being equal, there is more impetus.[82]

Thus Buridan's "impetus" seems quite similar to Galileo's "*impeto*," and Descartes' "*quantite de mouvement*."

Before going on to examine some of Buridan's applications of impetus theory, it is worth considering if, in fact, he has stated the principle of inertia or, equivalently, the conservation of momentum. After all impetus is a permanent quality, which is defined as the product of mass and velocity. Nevertheless to equate "impetus" and "momentum" would be a gross anachronism. It is not clear whether Buridan regards impetus as an effect of motion, as we might consider momentum, or as a cause of motion, which would make it similar to force. The quantitative definition would seem to argue for the former view. Buridan's use of impetus to explain projectile motion, and his association of impetus with motive power, would seem to favor the latter view. It seems most plausible to believe that Buridan, himself, was never quite sure of that distinction. Impetus is truly a giant step toward the principle of inertia and conservation of momentum but it is not equivalent to them.

Buridan's impetus also differs from Newton's momentum in that it is seen as also applying to circular motion. Thus, it includes some of what we would call both angular momentum and linear momentum. This is seen (as above) where Buridan states,

> Therefore, it seems to me that it ought to be said that the motor in moving a moving body impresses in it a certain impetus or a certain motive force of the moving body, [which impetus acts] in the direction toward which the mover was moving the moving body, either up or down, or laterally, or circularly.[80]

In addition, Buridan uses the rotation of a mill wheel to illustrate the permanence of impetus.

> And you have an experiment [to support this position]: If you cause a large and very heavy smith's mill [i.e., a wheel] to rotate and you then cease to move it, it will still move a while longer by this impetus it has acquired. Nay, you cannot immediately bring it to rest, but on account of the resistance from the gravity of the mill, the impetus would be continually diminished until the mill would cease to move. And if the mill would last forever without some dimunition or alteration of it, and there were no resistance corrupting the impetus, perhaps the mill would be moved perpetually by that impetus.[83]

There is further evidence that Buridan did not quite arrive at the inertial principle. In his discussion of the possible rotation of the earth, he considers the problem of why an arrow fired directly upward lands in the same spot.

> But the last appearance which Aristotle notes is more demonstrative in the question at hand. This is that an arrow projected from a bow directly upward falls again in the same spot of the earth from which it was projected. This would not be so if the earth were moved with such velocity. Rather before the arrow falls, the part of the earth from which the arrow was projected would be a league's distance away. But still the supporters would respond that it happens so because the air, moved with the earth, carries the arrow, although the arrow appears to us to move simply in a straight line because it is carried along with us. Therefore, we do not perceive that motion by which it is carried with the air. But this evasion is not sufficient because the violent impetus of the arrow in ascending would resist the lateral motion of the air so that it would not be moved as much as the air.[84]

Buridan uses impetus against the inertial solution. An inertial view would conclude that the arrow would share completely the motion of the earth and the air.[85] Buridan, rather, views the arrow as actually carried along by the air and that this lateral force of the air is resisted by the vertical impetus of the arrow. An inertial solution was, in fact, given by Nicholas Oresme, a follower of Buridan, which we will discuss in the next section.

It is surprising that Buridan does not believe in motion in a void.[86] It is the permanence of his impetus which would result in infinite motion in a void which causes him to reject this possibility. He rejects Aquinas' argument on the successive nature of such motion (*distantia terminorum*) by invoking God's power to produce instantaneous motion. He further shows the effect of the Edict of 1277 by stating that although God could produce motion in a void he could also prevent it. Buridan was aware of Avempace's arguments and seems to imply that if motion in a void could occur, which he denied, then Avempace's rather than Aristotle's views would be true.

> . . . in this eighth proposition and the reasons in support of it are based on the supposition that Avempace's opinion—–cited earlier—–is not true. However, I do not know how to disprove it, and indeed, agree more with it than with the opposite opinion. Now if this opinion of Avempace were conceded then this eighth proposition ought not to be conceded and Aristotle's supporting reasons are invalid.[87]

In the following discussion we shall deal with Buridan's use of impetus theory to explain the acceleration of falling bodies, and the continuation of this work by Albert of Saxony (1316-1390) and Nicholas Oresme. We also consider the differing views of these three scientists on whether the velocity of a freely falling object depends on the distance fallen or on the time of fall. These discussions are of particular importance since Galileo adopts similar, although not identical, views in his discussions of falling bodies. They also illustrate, once more, the important influence of medieval science on Galileo's work. We have argued previously that Galileo's early views included the idea of self-expending impetus. His change to a view of permanent impetus, which is particularly influenced by the discussion of projectile motion and falling bodies, is a crucial step in the development of his inertial views.

Buridan applied the idea of impetus to explain the acceleration of falling bodies as follows. A body falling only under the influence of gravity would, in accordance with Aristotelian principles, travel at a uniform speed. But along with this speed the body acquires an impetus which also moves it and this provides acceleration. Since the impetus continues to increase along with the speed the acceleration continues.

> From these [reasons] it follows that one must imagine that a heavy body not only acquires motion unto itself from its principal mover, i.e. its gravity, but that it also acquires unto itself a certain impetus with that motion. This impetus has the power of moving the heavy body in conjunction with the permanent natural gravity. And because that impetus is acquired in common with motion, hence the swifter the motion is, the greater and stronger

the impetus is. So, therefore, from the beginning the heavy body is moved by its natural gravity only; hence it is moved slowly. Afterwards it is moved by that same gravity and by the impetus acquired at the same time; consequently, it is moved more swiftly. And because the movement becomes swifter, therefore the impetus also becomes greater and stronger, and thus the heavy body is moved by its natural gravity and by that greater impetus simultaneously, and so it will again be moved faster; and thus it will always and continually be accelerated to the end.[88]

Buridan was unclear as to whether the velocity of fall depends on the time or the distance. At one point it seems he favors the latter:

Hence, throughout, the greater velocity does not arise from a greater proximity to the earth, or because the body has less air beneath it, but from the fact that the moving body is moved from a longer distance and through a larger space.[89]

At another point his meaning is not clear:

I conclude, therefore, that the accelerated natural movements of heavy and light bodies do not arise from greater proximity to [their] natural place, but from something else that is either near or far, but which is varied by reason of the length of the motion (*ratione longitudinis notus*).[90]

It is not obvious whether "length of motion" refers to time or distance.

We see that Buridan's impetus is similar to Avicenna's permanent "*mail*." His treatment of falling bodies is also similar to that of Abu'l-Barakat, although the earlier treatment does not contain permanent "*mail*," while Buridan's impetus is permanent. We have discussed earlier that this work was not available during this period, so it is likely that Buridan's work on falling bodies is a continuation of the work of other Scholastics.

We also note that Buridan applied impetus to explain the perpetual motions of the celestial bodies. (We shall discuss this in detail in the next section.)

Buridan's work was continued by both Albert of Saxony and Nicholas Oresme. Albert also uses the idea of increasing impetus with increasing speed to explain the acceleration of falling bodies. Like Buridan, he viewed impetus as permanent, although in the passage quoted below it is not clear whether the decrease of impetus is due to resistance or whether it is self-expending. Elsewhere, however, he used the example of a millstone to show the permanent nature of impetus just as Buridan had.

Albert used permanent impetus in treating the problem of dropping a stone through a hole in the earth, which, as we shall see, was also considered by Galileo.

According to this [theory], it would be said also that if the earth were completely perforated, and through that hole a heavy body were descending quite rapidly toward the center, that when the center of gravity of the descending body was at the center of the world, that body would be moved on still further [beyond the center] in the other direction, i.e. toward the heavens, because of the impetus in it not yet corrupted. And in so ascending, when the impetus would be spent, it would conversely descend. And in such a descent it would again acquire unto itself a certain small impetus by which it would be moved again past the center. When this impetus was spent, it would descend again. And so it would be moved, oscillating about the center until there no longer would be any such impetus in it, and then it would come to rest.[91]

Albert seems to believe that the velocity of fall is proportional to the distance travelled.

Therefore in the third conclusion it is understood that the speed is increased by double, triple, etc. in such a fashion that when some space has been traversed by this [motion], it has a certain velocity, and when a double space has been traversed by it, it is twice as fast, and when a triple space has been traversed by it, it is three times as fast, and so on.[92]

It is this view that Galileo seems to have adopted in his early work on falling bodies.[93]

Oresme states the correct view that the velocity of fall is proportional to time and, as we have discussed earlier, derived several of Galileo's results.

> That something is "continually accelerated" can be understood in two ways. In one way thus: An addition of velocity takes place by equal parts, or equivalently. For example, in this hour it is moved with some velocity, and in the second twice as quickly, and in the third three times as quickly.... Now as for the question at hand [of the acceleration of falling bodies], velocity in the motion of a heavy body increases in the first way and not the second.[94]

Oresme, however, believed that impetus was self-expending and that it was a function of both the acceleration and the velocity.[95] He also considers the question of an object dropped into a hole in the earth. He compares this motion to that of a pendulum. Obviously, Oresme did not prove, as we can, that these two motions are mathematically equivalent. He rather notes the similarity of the two kinds of periodic motion indicating the quality of his thinking.

> And this quality can be called "impetuosity." And it is not weight properly [speaking] because if a passage were pierced from here to the center of the earth or still further, and

something heavy were to descend in this passage or hole, when it arrived at the center it would pass on further and ascend by means of this accidental and acquired quality, and then it would descend again, going and coming several times in the way that a weight which hangs from a beam by a long cord [swings back and forth].[96]

A similar discussion appears in Galileo's "Dialogue Concerning the Two Chief World Systems."

Thus, if the earth were tunneled through the center, and a ball were let fall a hundred or a thousand yards toward the center, I verily believe that it would pass beyond the center and ascend as much as it descended. This is shown plainly in the experiment of a plummet hanging from a cord, which, removed from the perpendicular (its state of rest) and then set free, falls toward the perpendicular and goes the same distance beyond it——or only so much less as the cord, the resistance of the air, and other accidents impede it.[97]

I have put forth the observation of the pendulum so that you would understand that the impetus acquired in the descending arc, in which the motion is natural, is able by itself to drive the same ball upward by a forced motion through as much space in the ascending arc; by itself, that is, if all external impediments are removed.[98]

We see clearly the influence of the earlier work of Buridan, Oresme, and Albert of Saxony. We would, however, be doing Galileo an injustice if we were to say that these treatments are equivalent. Despite the marked similarity of the statements, Galileo's impetus does not contain the idea of motive force or a cause of motion that is included in the medieval view. His impetus is rather an effect and measure of motion. It is close to the idea of conservation of momentum and inertia, although as we shall see later Galileo does not arrive at a clear or correct statement of these principles. Nevertheless both his work and that of his medieval precursors are further steps toward the inertial principle.

We shall not deal extensively with the development of the work started by Buridan and the Merton College writers since there were no major developments until the sixteenth century, but we do note that this work was widely known throughout Europe.[99] Marsilius of Inghen and Henry of Hesse both extended Buridan's work by distinguishing between circular and rectilinear impetus. An interesting critique of impetus theory appears in the work of Blasius of Parma (died 1416).[100] In a discussion of a rebounding object Blasius rejects the explanation of impetus theory and rather speaks of the persistence of motion.

> And this is evident because when some heavy body meets a hard body it rebounds in a contrary direction since a quality such as motion cannot be destroyed instantaneously. It is true some say [the body rebounds] because of the impetus acquired in [the

> course of] the motion; but it is not helpful to
> speak this way in the present [context].[101]

Like Ockham, Blasius believes that motion will continue despite the absence of the force which originally produced it. It is yet another step toward the principle of inertia.

The spread of both Mertonian and Parisian physics continued throughout the fifteenth century. Impetus theory was discussed by Nicholas of Cusa (1401-1464). In *De Ludo Globi* he attributes the motion of the outer sphere to impetus. The Duke of Bavaria asks, "But how did God create the motion of the outermost sphere?" The German cardinal answers:

> Much as you give motion to the globe. But
> this sphere is not moved directly by God, the
> Creator, nor by the Spirit of God; as it is not
> you nor your spirit who move immediately
> the globe which is now rotating in front of
> you. It is, however, you who initiate this
> motion, since the impulsion of your hand,
> following your will, produced an impetus and
> as long as this impetus endures the globe
> continues to move.[102]

Cusa is clearly applying Buridan's idea of circular impetus to the heavens, just as Buridan had done. He also discussed the perpetual motion of an object set in motion on a perfectly smooth earth. This view, as we shall see, was later adopted by Galileo, who may have been influenced by Cusa.[103] We shall not deal with the physics contained in the notebooks of Leonardo da Vinci since it is not clear whether this had any influence on other scientists. He believed, however, that both the distance and velocity were proportional to time, which is an impossibility. "Therefore at each doubled quantity of time the length of the descent is doubled and also the swiftness of movement."[104] In 1555 Domingo de Soto applied the Merton College analysis and impetus theory explicitly to falling bodies.

> Movement uniformly nonuniform as to time is nonuniform in such a manner that if it is divided according to time (i.e., according to before and after), the middle point of any part at all exceeds [in velocity] the least velocity of that part by the same proportion that the mean is exceeded by the greatest velocity [of that part]. This species of movement belongs properly to things which are moved naturally and to projectiles.[105]

He goes on to state that if a body falling for an hour accelerated from a velocity of zero to a velocity of eight it would traverse the same distance as an object travelling uniformly for an hour with a speed of four, which is precisely the Merton College result.

Impetus theory was also discussed widely, particularly in Italy, during the sixteenth century. We have seen earlier that Galileo's early views were influenced by the theory of self-expending impetus and motion in a void of Philoponus and Avempace. Benedetti (1530-1590), who also subscribed to this theory, made another important criticism. He clearly distinguished between rectilinear and circular impetus. In his argument against the permanence of circular impetus he discusses the question of a stone thrown by a sling which moves tangentially, not circularly, after it is released.

> On the subject of this third type (of the potter's wheel) a problem may be phrased: why it is that, if a wheel of this kind parallel to the horizon rests upon one point and is as evenly [balanced] as possible, and if we revolve it with all our force and let it go it does not rotate forever.... Fourth, any portion of corporeal matter which moves by itself when an impetus has been impressed on it by any external motive force has a natural tendency to move in a rectilinear, not a curved path. And so if some portion of the circumference were separated off from the wheel in question, no doubt the separated part would move through the air in a straight line for some length of time. This we can understand by an example taken from slings used for throwing stones. In their case the impressed impetus of motion produces a rectilinear path by a certain natural propensity, when the stone shot out starts its rectilinear path; this path is along the straight line tangent to that circle which the stone previously described at the point at which it was let fly.[106]

Were it not for the phrase "for some length of time" Benedetti would have formulated the principle of inertia. He also recognized that the path of a projectile will not remain straight and attributes this both to the action of gravity and the decrease of impetus.

> Now it is true that the impressed impetus gradually and continuously decreases. Hence the downward tendency of the body, caused by its weight, enters at once and, mingling itself with the impressed force does not permit the line to remain straight for long, but causes it quickly to become curved.[107]

Galileo seems to favor the idea of the permanence of circular impetus only. Despite his views that motion along a horizontal plane would continue forever and that projectiles travel with constant horizontal speed, he regards these as approximations to motion parallel to the earth's surface. It seems that, although he was able to consider the absence of external resistance, he was never able to imagine the absence of gravity. This is made clear in his discussion of projectile motion. In his proof of the parabolic path of projectiles the objection is raised that the horizontal motion cannot be constant and Galileo admits that this is true.

> A projectile which is carried by a uniform horizontal motion compounded with a naturally accelerated vertical motion describes a path which is a semi-parabola.[108]

> Simplicio: To these difficulties, I may add others. One of these is that we suppose the horizontal plane, which slopes neither up nor

down, to be represented by a straight line as if each point on this line were equally distant from the center, which is not the case; for as one starts from the middle [of the line] and goes toward either end, he departs farther and farther from the center [of the earth] and is constantly going uphill. Whence it follows that the motion cannot remain uniform through any distance whatever, but must continually diminish. . . .

Salviati: All these difficulties and objectives which you urge are so well founded that it is impossible to remove them; and as for me, I am ready to admit them all, which indeed I think our Author would also do. I grant that these conclusions proved in the abstract will be different when applied in the concrete and will be fallacious to this extent, that neither will the horizontal motion be uniform. . . . Some consider this assumption [constant horizontal motion] permissible because, in practice, our instruments and the distances involved are so small in comparison with the enormous distance from the center of the earth that we may consider a minute of arc on a great circle as a straight line.[109]

A similar view is indicated in his discussion of motion on an inclined plane. Galileo regards only the surface of the earth as horizontal. "Hence, along the horizontal, by which we understand a surface, every point of which is equidistant from this same common center, the body will have no momentum whatever."[110] We shall discuss Galileo's work on inclined planes and their relevance to his inertial views in a later section.

Galileo's inability to abstract from the effects of gravity is further illustrated in his discussion of whether or not objects would be extruded from a rotating earth. In a discussion similar to the discussion of projectiles given by Benedetti, Galileo states,

> ... the circular motion of the projector impresses an impetus upon the projectile to move, when they separate, along the straight line tangent to the circle of motion at the point of separation, and that continuing with this motion, it travels ever farther from the thrower. And you have said that the projectile would continue to move along that line if it were not inclined downward by its own weight from which fact the line of motion derives its curvature. It seems to me that you also knew by yourself that this bending always tends toward the center of the earth, for all heavy bodies tend that way.[111]

Were it not for the last sentence Galileo's statement would be in exact accord with the principle of inertia. This last sentence, in which Galileo indicates that objects *always* tend toward the center of the earth, illustrates his view of the pervasiveness of gravity. Nowhere, in fact, does Galileo ever state what the motion of a projectile would be if no forces were acting on it. Galileo's view is different from that of Benedetti given above. Benedetti misses the inertial principle because he views impetus as self-expending. Galileo regards impetus as permanent but is unable to neglect the effects of gravity.

Galileo's view on the permanence of circular motion is indicated in the following statement. We see again his inability to neglect gravity.

> Salviati: Then in order for a surface to be neither downward or upward, all parts must be equally distant from the center. Are there any such surfaces in the world?

> Simplicio: Plenty of them; such would be the surface of our terrestrial globe if it were smooth, and not rough and mountainous as it is. But there is that of the water, when it is placid and tranquil.

> Salviati: Then a ship when it moves over a calm sea, is one of these movables which courses over a surface that is tilted neither up nor down, and if all external and accidental obstacles were removed, it would thus be disposed to move incessantly and uniformly from an impulse once received.[112]

Galileo's view seems quite similar to that of Nicholas of Cusa discussed earlier.[113] Galileo has come very close to the principle of inertia but he has not stated it exactly.

In this section we have seen that Galileo's views on inertia, as evidenced in his treatment of projectiles and falling bodies, were influenced by the kinematic treatment of motion of William of Ockham and the Merton College scientists and the permanent impetus theory of Buridan, and his followers. It seems clear that these ideas, combined with Galileo's never-relinquished earlier view, following Philoponus and Avempace, that both the action and resistance of a medium are unnecessary for motion, are crucial steps in the development of an inertial mechanics.

The works of these medieval scientists were quite available at the time of Galileo. No fewer than seventeen printed versions of the Merton College Mean Speed Theorem were available[114] and the writings of Buridan, Oresme, and Albert of Saxony were readily available either directly or indirectly. In particular, the University of Padua, where Galileo worked after his stay at Pisa, contained the works of Heytesbury, Swineshead, Dumbleton, Buridan, Albert of Saxony, and others.[115] Galileo's early notes also include citations of Albert of Saxony, Hentisberus (Heytesbury), and the "Calculator" (Swineshead).[116]

In the next two sections we will examine the influence on Galileo's inertial views of the medieval discussions of the possible rotation of the earth and of motion on an inclined plane.

FOUR

HEAVEN AND EARTH; THE POSSIBLE ROTATION OF THE EARTH

Two other medieval discussions are important to the history of the principle of inertia. First, since we have seen that perpetual uniform circular motion was allowable for heavenly bodies, any breakdown of the hierarchical distinction between celestial and terrestrial motion could imply that perpetual motion could occur on earth. Second are the discussions of the possible rotation of the earth. Since we may regard the principle of inertia as stating the equivalence of uniform rectilinear motion and rest, the discussions on the relativity of motion, particularly those concerned with the possible rotation of the earth, may be considered steps toward that principle.[117] In addition, several of the arguments used to support the idea of the earth's rotation implicitly involve inertial ideas.

The ancient Greeks were unanimous in regarding the motion of the heavens as uniform circular motion. It was generally accepted that the earth was fixed and immobile in the center of the universe. This view was disputed by Philolaus, a Pythagorean, who believed that the earth and all other heavenly bodies, including the sun, circled a central fire. Heraclides of Pontus and Aristarchus both hypothesized the diurnal rotation of the earth.

> There have been some, among them Heraclides of Pontus and Aristarchus, who thought that the phenomena could be accounted for by supposing the heavens and stars to be at rest, and the earth to be in motion about the poles of the equator from west [to east].[118]

Heraclides extends this view further to include the motion of Venus and Mercury about the sun, while Aristarchus evolved a complete heliocentric theory. It is only the medieval discussions of the rotation of the earth that are important for our discussion.

We shall deal first with the breakdown of the distinction between terrestrial and celestial motions. We have seen earlier that Avempace used the finite motion of the heavenly bodies to support his view that motion in a void would not be instantaneous. He further tries to apply the mechanics of earthly bodies to the intelligences or spirits that move the heavenly spheres. Just as the motion of an object on earth is determined by the differences between force and resistance, so it is analogously in the heavens, where we consider the difference in perfection between the mover and the thing moved.

And this is only due to a difference in perfection between the mover and the thing moved.
Therefore, when the mover is of greater
perfection, that which is moved by it will be
swifter; and when the mover is of less perfection it will be nearer [in perfection] to the
thing moved and the motion will be
slower.[119]

Aquinas, too, uses the example of the heavenly bodies
to show motion with finite velocity in a void. The
breakdown was furthered by the Edict of Paris in 1277
where the heavens could undergo rectilinear motion,
which had previously been restricted to terrestrial
objects.

Franciscus de Marchia applied his theory of self-
expending impressed force not only to projectiles and
other motions on earth but also, as we have seen, to the
heavenly bodies.

From this it follows that with the intelligences ceasing to move the heavens, the
heavens would still be moved, or revolve, for a
time by means of a force of this kind following and continuing the circular motion, as
is evident in a potter's wheel which revolves
for a time after the first mover has ceased to
move [it].[120]

Both Avempace and Franciscus de Marchia had applied
dynamics to the actions of the intelligences which move
the heavenly bodies. These spirits were exorcised in the

work of Buridan and Albert of Saxony who apply permanent impetus to the motions of heavenly bodies. Buridan states:

> And thus we could imagine that it is unnecessary to posit intelligences as the movers of celestial bodies since the Holy Scriptures do not inform us that intelligences must be posited. For it could be said that when God created the celestial spheres, He began to move each of them as He wished, and they are still moved by the impetus He gave to them because, there being no resistance, the impetus is neither corrupted nor diminished.[121]

It is, of course, a long way from these statements to Newton's Law of Universal Gravitation, but they do constitute a first step toward such a universal mechanics. In addition, by breaking down the distinction between celestial and terrestrial motion they allow for the possibility of perpetual motion on earth and may be considered a step toward the principle of inertia.

The rotation of the earth was also discussed during the Middle Ages. Simplicius' commentary on "On the Heavens" which discussed the theories of Heraclides and Aristarchus was translated in 1271. The question was also dealt with by Arabic and Indian writers. It was also mentioned by several medieval writers, including St. Thomas Aquinas, who rejected the idea. The earliest medieval reference I have found is by an anonymous late-twelfth-century writer who states, "For no one denies but that it [the earth] may well revolve on its own axis."[122] This writer not only discusses the question of the earth's rotation but seems to regard it as the majority view. The major contributions to this discussion were made in the fourteenth century by Buridan and Oresme. We shall not deal with all of their arguments on the rotation of the earth, many of which are also repeated by Copernicus and Galileo, but only with those which have inertial implications. In these discussions we shall also see the breakdown of the distinction between celestial and terrestrial motions.

The discussions of the relativity of motion were most important. This is, as we have discussed earlier, an important step toward the principle of inertia. It is quite instructive to look at the views on this subject by Buridan, Oresme, Copernicus, and Galileo simultaneously so that the similarity of their views will be apparent.

> Buridan: It should be shown that many people have held as probable that it is not contradictory to appearances for the earth to be moved circularly in the aforesaid manner, and that on any given natural day it makes a complete rotation from west to east by returning again to the west——that is, if some part of the earth were designated [as the part to observe]. Then it is necessary to posit that the stellar sphere would be at rest, and then night and day would take place through such a motion of the earth, so that that motion of the earth would be a diurnal motion. The

following is an example of this: If anyone is moved in a ship and he imagines that he is at rest, then, should he see another ship which is truly at rest, it will appear to him that the other ship is moved. This is so because his eye would be completely in the same relationship to the other ship regardless of whether his own ship is at rest and the other moved, or the contrary situation prevailed. And so we also posit that the sphere of the sun is everywhere at rest and the earth in carrying us would be rotated. Since, however, we imagine that we are at rest, just as the man located on the ship which is moving swiftly does not perceive his own motion nor the motion of the ship, then it is certain that the sun would appear to us to rise and then to set, just as it does when it is moved and we are at rest. . . . It is undoubtedly true that, if the situation were just as this position posits, all celestial phenomena would appear to us just as they now appear.[1 2 3]

Oresme: Now, I take it as a fact, that local motion can be percieved only if we can see that one body assumes a different position relative to another body. For example, if a man is in boat A, which is moving very smoothly either at rapid or slow speed, and if this man sees nothing except another boat B, which moves precisely like boat A, the one in which he is standing, I maintain that to this man it will appear that neither boat is moving. If A rests while B moves, he will be aware that B is moving; if A moves and B rests, it will seem to the man in A that A is resting and B is moving, just as before. Thus, if A rested an hour and B moved, and during the next hour it happened conversely that A moved and B rested, this man would not be able to sense this change or variation; it would seem to him that all this time B was moving. This fact is evident from experience, and the reason is that the two bodies A and B have a continual relationship to each other so that, when A moves, B rests and conversely when B moves, A rests. It is stated in Book Four of "The Perspective" by Witelo that we do not perceive motion unless we notice that one body is in the process of assuming a different position relative to another.[124] I say, therefore, that if the higher of the two parts of the world mentioned above were moved today in daily motion——as it is——and the lower part remained motionless and if tomorrow the contrary were to happen so that the lower part moved in daily motion and the higher——that is, the heavens, etc.——remained at rest, we should not be able to sense or perceive this change.[125]

Copernicus: For every apparent change in place occurs on account of the movement either of the thing or of the spectator, or on account of the necessarily unequal movement of both. For no movement is perceptible relatively to things moved equally in the same directions——I mean relatively to the thing seen and the spectator. Now it is from the Earth that the celestial circuit is beheld and presented to our sight. Therefore, if some movement should belong to the Earth it will appear, in parts of the universe which are outside, as the same movement but in the opposite direction, as though the things outside were passing over. And the daily revolution in especial is such a movement.[126]

Galileo: Salviati: It is obvious, then, that motion which is common to many moving things is idle and inconsequential to the relation of these movable among themselves, nothing being changed among them, and that it is operative only in relation that they have with other bodies lacking that motion, among which their location is changed. Now, having divided the universe into two parts, one of which is necessarily movable and the other motionless, it is the same thing to make the earth alone move, and to move all the rest of the universe, so far as concerns any result which may depend upon such movement. For the action of such a movement is only in the relation between celestial bodies and the earth, which relation alone is changed.[127]

We see that the four writers hold the same view that motion is relative, and thus that Galilean relativity was derived in the fourteenth century. Oresme and Buridan were no doubt influenced by Ockham's view of motion and by other medieval writers, notably Witelo and Alhazen (see note 124).

The arguments given above are not absolutely correct since technically it is only the relativity of uniform rectilinear motion which is equivalent to the principle of inertia. The authors do not distinguish between this kind of motion and uniform circular motion, which is accelerated and thus detectable in an absolute sense. We can, for example, detect the rotation of the earth by observing the change in the plane of oscillation of a pendulum. Nevertheless the general idea of the relativity of motion indicated above is an important step toward the inertial principle, despite the fact that the arguments are not completely valid.

The inability to detect uniform rectilinear motion is illustrated explicitly in the following quotations from Oresme, Copernicus, and Galileo. All three writers consider motion in a closed mechanical system and argue that the motion of such a system is undetectable. This is in agreement with the principle of inertia which regards rest and uniform rectilinear motion as equivalent.

> Oresme: Inside the boat moved rapidly east-ward, there can be all kinds of movements—— horizontal, criss-cross, upward, downward, in all directions——and they seem to be exactly the same as those when the ship is at rest.[128]

Copernicus: As a matter of fact, when a ship floats on over a tranquil sea, all the things outside seem to the voyagers to be moving in a movement which is the image of their own, and they think on the contrary that they themselves and all the things with them are at rest.[129]

Galileo: Shut yourself up with some friend in the main cabin below decks on some large ship, and have with you there some flies, butterflies, and other small flying animals. Have a large bowl of water with some fish in it; hang up a bottle that empties drop by drop into a wide vessel beneath. With the ship standing still, observe carefully how the little animals fly with equal speed to all sides of the cabin. The fish swim indifferently in all directions; the drops fall into the vessel beneath; and, in throwing something to your friend, you need throw it no more strongly in one direction than another, the distances being equal; jumping with your feet together you pass equal spaces in every direction. When you have observed all these things carefully, have the ship proceed with any speed you like, so long as the motion is uniform and not fluctuating this way and that. You will discover not the least change in all the effects named, nor could you tell from any of them whether the ship was moving or standing still.[130]

The inertial views of these writers are further illustrated in their answers to the objection that if the earth rotated a constant wind would blow from east to west. They all agree that the air and water share the motion of the earth just as the objects on a ship share its motion. Their views are given below.

Buridan: If anyone were moving very swiftly on horseback he would feel the air resisting him. Therefore, similarly with the very swift motion of the earth in motion, we ought to feel the air noticeably resisting us. But there respond that the earth, the water, and the air in the lower region are moved simultaneously with diurnal motion. Consequently there is no air resisting us.[131]

Oresme: To the second experience, the reply seems to be that, according to this opinion, not only the earth moves, but also with it the water and the air, as we stated above, although the water and air here below may be moved by the winds or other forces. In a similar manner, if the air were closed in on a moving boat, it would seem to a person in that air that it was not moving.[132]

Copernicus: Then what would we say about the clouds and the other things floating in the air or falling or rising up, except that not only the Earth and the watery element with which it is conjoined are moved in this way but also

no small part of the air and whatever other
things have a similar kinship with the Earth?
Whether because the neighboring air, which is
mixed with earthly and watery matter, obeys
the same nature as the Earth or because the
movement of the air is an acquired one, in
which it participates without resistance on
account of the contiguity and perpetual rotation of the Earth. . . . Hence the air which is
nearest the earth and the things floating in it
will appear tranquil, unless they are driven to
and fro by the wind or some other force, as
happens. For how is the wind in the air
different from a current in the sea.[133]

Galileo: I might add that at least that part of
the air which is lower than the highest moun-
tains must be swept along and carried around
by the roughness of the earth's surface, or
must naturally follow the diurnal motion
because of being a mixture of various terrestrial vapors and exhalations.[134]

It seems that on this question the views of Buridan
and Oresme are closer to the principle of inertia than
those of Copernicus or Galileo. Buridan and Oresme
state that the air and water completely share the motion
of the earth while Galileo seems to believe that the air is
pushed along by the earth's surface. Galileo and Copernicus also seem to regard the motion of the air as due
possibly to a mixture of earth and water in the air.
Nevertheless when we consider the views as a whole

they are inertial.

Buridan, however, did not believe in the rotation of the earth. As we have seen earlier, he rejects this view in his discussion of the problem of the arrow shot straight up. The Aristotelian arguments had stated that if the earth were rotating the arrow would not land in the same place. Buridan, as we have seen, believes that the air would have to carry the arrow along rather than the inertial view that the arrow would share the motion of the earth. Thus the upward impetus of the arrow will resist the lateral motion of the air and it would not be moved with the air and it would not fall in the same place. Impetus, which is an important step toward the principle of inertia, is here used to deny it.

> But the last appearance which Aristotle notes is more demonstrative in the question at hand. This is that an arrow projected from a bow directly upward falls again in the same spot of the earth from which it was projected. This would not be so if the earth were moved with such velocity. Rather before the arrow falls, the part of the earth from which the arrow was projected would be a league's distance away. But still the supporters would respond that it happens so because the air, moved with the earth, carries the arrow, although the arrow appears to us to be moved simply in a straight line motion because it is being carried along with us. Therefore, we do not perceive that motion by which it is carried with the air. But this evasion is not

sufficient because the violent impetus of the
arrow in ascending would resist the lateral
motion of the air so that it would not be
moved as much as the air.[135]

Oresme and Galileo, on the other hand, accept the
inertial view that the projectile shares the motion of the
earth.

Oresme: Concerning the third experience,
which seems more complicated and which
deals with the case of an arrow or stone
thrown up in the air, etc. one might say that
the arrow shot upward is moved toward the
east very rapidly with the air through which it
passes, along with all the lower portion of the
world which we have already defined and
which moves with daily motion; for this
reason the arrow falls back to the place from
which it was shot into the air. Such a thing
could be possible in this way, for, if a man
were in a ship moving rapidly eastward without his being aware of the movement and if
he drew his hand in a straight line down along
the ship's mast, it would seem to him that his
hand were moving with a rectilinear motion;
so, according to this theory it seems to us that
the same thing happens with the arrow which
is shot straight down or straight up.[136]

Galileo: In replying to this, those who make
the earth moveable answer that the cannon

and the ball which are on the earth share its motion, or rather that all of them together have the same motion naturally. Therefore the ball does not start from rest at all, but to its motion about the center joins one of projection upward which neither removes nor impedes the former. In such a way, following the general eastward motion of the earth, it keeps itself continually over the gun during both its rise and its return. You will see the same thing happen by making the experiment on a ship with a ball thrown perpendicularly upward from a catapault. It will return to the same place whether the ship is moving or standing still.[137]

We again note that strictly speaking the discussion given above on both the motion of the air and the motion of projectiles is incorrect. Only rectilinear motion can be shared. Benedetti and Galileo have already noted this when they stated that a stone launched from a string proceeds tangentially. Thus, the earth cannot give circular motion to either the air or to a projectile. The illustrations of the motion of air and projectiles on a uniformly moving ship are correct and indicate that these writers did have a fair understanding of the inertial principle.[138]

It will come as a shock that, despite his many excellent arguments in favor of the rotation of the earth, Oresme ultimately rejects the idea. Although he had previously argued that the Bible speaks in everyday language in discussing astronomical phenomena, as

Galileo would do later, it is finally the Scriptural argument that is decisive for Oresme.

> However, everyone maintains, and I think myself, that the heavens do move and not the earth: For God hath established the world which shall not be moved, in spite of contrary reasons because they are clearly not conclusive persuasions.[139]

This conclusion is not due to fear or to any external pressure from the Church, but seems rather the sincere belief of a devout man. Oresme was Bishop of Lisieux and is here defending his faith against attacks by reason. He is arguing that if reason cannot decide a relatively simple physical question like the rotation of the earth, it cannot deal with theological questions.[140] We should not, on this account, deny the importance of Oresme's discussions of the rotation of the earth and their influence on the history of inertia.

In this section we have seen the importance of the medieval discussions of motions in the heavens and on earth, for the development of the principle of inertia. We have seen that a breakdown in the distinction between the heavens and the earth can lead to the idea of perpetual motion on earth. The influence of these discussions is seen in the consideration of the rotation of the earth where the same motion is attributed to either the earth or the heavens.

The arguments of Buridan, Oresme, Copernicus, and Galileo on this last question are so similar that it would be surprising if there were no connection among them.

Buridan and Oresme were both at the University of Paris at the same time and it is likely that Oresme was a student of Buridan, so we would expect similar views. There is no difficulty in relating Copernicus to Galileo since Galileo makes it clear that he has read Copernicus. We have seen earlier that the medieval work was available directly to Galileo. The only difficulty is in finding a connection between the medieval scientists and Copernicus. Copernicus never refers to any medieval sources. It is known that Parisian mechanics was known in Cracow and eastern Europe as well as in Italy.[141] Since Copernicus studied in both Cracow and Italy it is clear that this work was available to him. Given the similarity of the discussions of Copernicus as compared with those of Buridan and Oresme it seems reasonable to assume that he was familiar with their work. Thus the medieval discussions of the rotation of the earth, which we have indicated earlier are inertial in character, form an important step in the development of the principle of inertia.

FIVE

MOTION ON AN INCLINED PLANE

Galileo's views on inertia were no doubt influenced by his discussion of motion on an inclined plane. It is by now commonplace to consider that Galileo stated the principle of inertia by noting that motion downward on an inclined plane is accelerated and that motion upward on such a plane is decelerated. Thus motion on a horizontal plane, which is neither accelerated nor decelerated, would be perpetual. We shall see that Galileo did eventually arrive at such a view. However, he regarded this as an approximation to perpetual circular motion. (See quotation in Chapter Three and below.) Nevertheless it is still an important step toward the correct statement of the principle of inertia.

Galileo's treatment of motion on an inclined plane seems to owe much to the discussions of statics in Greek, Arabic, and medieval sources. The detailed study of this influence is beyond the scope of this paper but we will give a brief sketch of some of the early works which deal with motion on an inclined plane.[142]

The question of motion on an inclined plane seems to have been considered first by Hero of Alexandria (first century B.C.) who states that no force is necessary to move an object on a smooth horizontal surface. He first considers the motion of objects on planes inclined in different directions and then considers the flow of water on inclines.

There are those that think burdens lying flat are moved by an equal power [only], wherein they hold wrong opinions. So let us explain that burdens placed in the way described are moved by a power smaller than any known power, and we shall explain why this is not evident in practice. Let us imagine a burden lying flat, and let it be regular smooth and let its parts be coherent with each other. And let the surface on which the burden lies be able to be inclined to both sides, I mean to the right and the left. And let it be inclined first towards the right. Then it is evident to us that the supposed burden will incline toward the right side because the nature of burdens is to move downwards, if nothing holds them and hinders them from the movement; and again if the inclined side is lifted to a horizontal position and comes into equilibrium, the burden will come to rest in this position. And if it is inclined to the other side, I mean to the left side, the burden will again sink towards the inclined side, even if the inclination is very small, and the burden will need no power to move it, but will need a power to hold it so it does not move. And if the burden is again in equilibrium without inclination to any side then it will stay there although there is no power to hold it, and it will not cease being at rest, until the surface inclines to one side or another. Then it will incline to that side. And the burden that is ready to go to every side, how can it fail to need to move it a very small power of the size of the power that will incline it? And so the burden is moved by any small power. . . .

The waters that are on a surface that is not inclined will not flow, but remain without inclining to either side. But if the slightest inclination comes to the water, then all of it will incline towards that side.[143]

This is obviously not an inertial view of motion since nothing is really said about what would happen to an object that was set in motion on such a plane, but it is certainly a beginning. We shall see later that Galileo adopted a similar view in his early work.

Hero's work does not seem to have been known directly in the Middle Ages but the study of statics begun in the works of pseudo-Aristotle (*Mechanica*),[144] Archimedes, Pappus, and other Greek and Arabic writers is continued by Jordanus of Nemore (fl. 1230-1260) and his followers in the thirteenth century. Jordanus' work on statics is quite extensive but for the purposes of this paper we shall consider only some of his postulates on motion.

R1.001. The movement of every weight is toward the center (of the world), and its force is a power of tending downward and of resisting movement in the contrary direction.

R1.002. That which is heavier descends more quickly.

R1.003. It is heavier in descending, to the degree that its movement toward the center (of the world) is more direct.

R1.004. It is heavier in position when in that position its path of descent is less oblique.[145]

It is from these postulates that Cardano and Benedetti seem to derive the idea that only a minimal force is necessary to move an ideal sphere on a perfect horizontal plane.[146] Cardano attempted to prove that the effective weight of a body which is on an inclined plane depends on the angle of the plane. This is quite similar to axiom R1.004 of Jordanus given above. In connection with this proof he remarks that it is common knowledge that only a minimal force is required to move a sphere on a horizontal plane.[147]

Benedetti arrived at the same conclusion in his commentary on the mechanics of pseudo-Aristotle.

It is clear to us that we shall pull it (a sphere) without any difficulty or resistance because in such a case the center will never change its position.[148]

In the proof of this statement Benedetti seems to adhere directly to the postulates of Jordanus. In connection with this proof Benedetti remarks that the plane is only an approximation to the spherical surface of the earth.

> And note, in this connection, that even if I call it a plane, yet I do not wish it understood as a perfect plane, but as a perfectly spherical surface around the center that is the goal of [freely falling] heavy bodies. For by reason of the great extent of such a surface we shall be able to picture no noteworthy difference between a perfect plane of small extent and the [infinitesimal] curvature of such a spherical surface.[149]

We have seen earlier that Galileo adopted a similar view and Benedetti's work may have had an influence on him.

Galileo's earliest work on the inclined plane also seems to indicate the influence of Jordanus. He assumes only one principle:

> ... that a heavy body tends to move downward with as much force as is necessary to lift it up; i.e., it tends to move downward with the same force with which it resists rising.[150]

This is quite similar to Jordanus' first postulate given above. He then goes on to state that motion on a horizontal plane requires no force.

... then any body on a plane parallel to the horizon will be moved by the very smallest force, indeed, by a force less than any given force.... A body subject to no external resistance on a plane sloping no matter how little below the horizon will move down [the plane] in natural motion, without the application of any external force. This can be seen in the case of water. And the same body on a plane sloping upward, no matter how little, above the horizon, does not move up [the plane] except by force.[151]

Galileo's discussion seems quite similar to that of Hero, and he too does not discuss what would happen once the object was set in motion.[152] This same view is also stated in his later work "On Mechanics." After a discussion of the flow of water downhill, the consideration of the motion of a hard polished ball on a hard surface or a frozen pond, he concludes:

Hence it is perfectly clear that on an exactly balanced surface the ball would remain indifferent and questioning between motion and rest, so that any the least force would be sufficient to move it, just as on the other hand any little resistance, such as that merely of the air that surrounds it, would be capable of holding it still.[153]

Galileo does seem to have considered the subsequent motion of such a body quite early in his career although he does not publish this view until later. In April 1607 one of Galileo's students, Benedetto Castelli, wrote to him as follows:

> ... from Your Excellency's doctrine. That although to start motion the mover is necessary, yet to continue it the absence of opposition is sufficient.[153]

Galileo seems to believe that such motion would be perpetual. In his "Letters on Sunspots" he remarks:

> And therefore, all external impediments removed, a heavy body ... will maintain itself in that state in which it has once been placed; that is if placed in a state of rest, it will conserve that; and if placed in movement toward the west, it will maintain itself in that movement.[154]

This is not, however, a statement of the principle of inertia since the motion takes place on a spherical surface concentric with the earth and again illustrates Galileo's view on the persistence of circular motion, and his inability to neglect the effects of gravity.

Galileo's views are further developed in the "Dialogues Concerning Two New Sciences" published in 1638.

> Furthermore we may remark that any velocity once imparted to a moving body will be rigidly maintained as long as the external

causes of acceleration or retardation are removed, a condition which is found only on horizontal planes; for in the case of planes which slope downwards there is already present a cause of acceleration, while on planes sloping upward there is retardation; from this it follows that motion along a horizontal plane is perpetual; for, if the velocity is uniform, it cannot be diminished or slackened, much less destroyed.[155]

Here then is the proof of the contention that Galileo stated the inertial principle. Indeed it would seem so, were it not for the fact that Galileo has previously defined horizontal as a surface equidistant from the center of the earth. (See quotation in Chapter Three.) Galileo has again come very close to the inertial principle but he has not stated it absolutely correctly. We have seen in this section another indication of an influence of medieval and ancient thought in Galileo's ideas.[156]

SIX

GALILEO AND BEYOND

The problem of inertial motion was also discussed by several of Galileo's contemporaries. It is still a matter of debate as to which one of them was the first to publish the correct statement of the principle of inertia.

Pierre Gassendi (1592-1655) has an excellent claim to rank among the first to state this principle. He seems to have been the first scientist to perform publicly and publish the results of an experiment to see whether a stone dropped from the mast of a moving ship will fall parallel to the mast or lag behind.[157] Galileo, of course, had discussed such an experiment in his defense of the Copernican system. Gassendi arrives at his inertial view from an incorrect concept of gravity. He could not envision action at a distance and regarded the attraction of gravity as due to an effluvium or emanation of particles from the attracting body. He states,

> [Let us] conceive a stone in those imaginary
> spaces that are extended beyond this world,

and in which God could create other worlds; do you think that, at the very moment that it [the stone] would be formed there, it would fly towards the Earth and not, rather remained unmoved where it was first placed, as if, so to say, it had no *up*, nor *down*, where it should tend and where from it should withdraw?

But if you think that it will come here, imagine that not only the Earth but the whole world is reduced to nothing, and that these spaces are completely empty as [they were] before God created the world; then, indeed, as there will be no center, all spaces will be similar; it is obvious that the stone will not come here, but will remain motionless in its place. Now let the world, and in it the Earth be put back again: will the stone immediately drive here? If you say that it will, it is necessary [to admit it] that the Earth will be felt by the stone, and therefore the Earth must transmit to it a certain force, and send out the corpuscles, by which it gives to it an impression of itself, in order, so to say, to announce to it that it [the Earth] is restored in being and put back in the same place. How, otherwise, could you understand that the stone should tend towards the Earth?[158]

Thus if God can create a void then there could be no influence of the earth on a body in such a void. This view of the void seems similar to that expressed in the

Edict of Paris of 1277. The discussion of the possibility of God creating other worlds is similar to another of the statements in the Edict in which the thesis "That the First cause could not make several worlds"[159] is condemned. If gravity cannot act on a body through an imaginary void then once a body is in motion, the motion will continue indefinitely in the same direction. Gassendi goes even further and asserts that this is true not only in a void but on earth.

> All that has no other aim than to make us understand that motion impressed [on a body] through void space where nothing either attracts, or resists will be uniform, and perpetual; and that, therefrom, we conclude that all motion that is impressed on a body is in itself of that kind; so that in whatever direction you throw a stone, if you suppose that, at the moment in which it leaves the hand, by divine power, everything besides this stone is reduced to nothing, it would result that the stone will continue its motion per-petually and in the same direction in which the hand has directed it. And if it does not do so [in fact], it seems that the cause is the admixion of the perpendicular motion which intervenes because of the attraction of the Earth, which makes it deviate from its path (and does not cease until it arrives at the Earth), just as iron scrapings thrown near a magnet do not move in a straight line but are deviated toward the magnet.[160]

Thus Gassendi has given, at the very least, a close approximation to the principle of inertia. He still does not regard motion as a state rather than as a process which is required for a truly inertial view. We have discussed this same distinction earlier in our comparison of Buridan's "impetus" and Galileo's "impeto." Galileo does not make this distinction explicitly but it does seem that he viewed motion as a state.

The modern inertial view is stated explicitly by Rene Descartes (1596-1650). Descartes first discussed this in his incomplete work "Le Monde" which was started around 1630 but withdrawn after the condemnation of Galileo in 1633. This work was published in 1662 but virtually identical views are expressed in Descartes' "Principles of Philosophy" published in 1644. Descartes states quite clearly that motion and rest are equivalent states, which we have seen is the correct inertial view.

> This motion has not a higher degree of reality than rest; quite the contrary, I conceive that rest is just as much a quality that has to be attributed to matter when it remains in a place as motion is one that has to be attributed to matter when it changes place.[161]

Descartes' definition of motion is reminiscent of the medieval views on the relativity of motion which we have seen earlier in the work of Ockham, Buridan, and Oresme.

> But if we consider what must be understood as motion, not according to the vulgar use [of

the term] but according to the truth [*ex rei
veritate*], in order to attribute to it a deter-
minate nature, we shall say that it is a translation of a part of matter, or of a body, from
the vicinity of those bodies that immediately
touch it, and that are considered at rest, to
the vicinity of others.[162]

This is further illustrated by his view that motion is
completely reciprocal.

We cannot conceive that a body AB could be
transported from the vicinity of a body CD
without knowing also that the body CD is
transported from the vicinity of the body AB,
and that there is needed, obviously, as much
force and action for the one as for the other.
Thus, if we should want to attribute to
motion a proper nature, not related to anything else, we would say that, when two
reciprocally contiguous bodies become
mutually separated, and transported one to
one side and the other to the other, there will
be as much motion in the one as in the
other.[163]

Descartes combined this view of motion with the idea
that the immutability of God implies that He acts in an
immutable manner and conserves the quantity of
motion in the world.

We understand also that it is a perfection in
God that not only is He immutable in

Himself, but also that He acts in a most constant and immutable manner. . . . Wherefrom it follows that it is in the highest degree comformable to reason to think . . . that He maintains in matter the same quantity of motion with which He created it.[164]

From this, Descartes derives several laws of nature,

. . . of which the first is that everything, considered as simple and undivided perseveres, as far as it can, in the same state and never changes [its state] but for external causes. Thus, if a certain part of matter is square, we are convinced very easily that it will perpetually remain square. . . . If it is at rest, we do not believe that it will ever begin to move if not compelled by some cause. Nor is there any reason to think that, if it moves. . . and is not impeded by anything, it should ever cease to move with the same force. It is therefore to be concluded that a thing which moves, will move forever as far as it can [and will not tend to rest because] rest is contrary to motion, and nothing by its own nature can tend toward its contrary, that is toward its own destruction; [therefrom comes] the first law of nature, that everything, as much as in it lies, perseveres always in the same state; thus that everything which once started to move will continue to move forever.[165]

This is, in fact, the proper statement of the principle of inertia. Descartes has turned the problem around. Rather than asking why motion continues he asks why it should stop.

> For, having supposed the preceding Rule, we are exempt from the trouble in which the *Docti* find themselves when they want to give a reason why a stone continues to move for some time after having left the hand of the one who threw it: for one should rather ask why it does not continue to move forever.[166]

We have thus returned to Aristotle's original question and even to his original answer (see Chapter One). The difference is that Aristotle regards the answer as absurd while Descartes regards it as true.

Newton was familiar with the work of Descartes, Galileo, and Gassendi and, in fact, attributes the discovery of the principle of inertia to Galileo. Nevertheless it is clear that Newton had studied the works of Descartes quite thoroughly.

We now come to the end of our history with the statement of the principle of inertia or Newton's "First Law of Motion."

> Every body continues in its state of rest, or of uniform motion in a right line, unless it is compelled to change that state by forces impressed upon it.[167]

CONCLUSION

It is to be hoped that this paper has demonstrated the obvious conclusion that the discussions of medieval scientists form an essential link between the anti-inertial views of Aristotle and the almost inertial views of Galileo, and through him to the inertial principle of Descartes and Newton. It would, in fact, have been more shocking if such a link did not exist. We might rather have turned the question around and asked not whether medieval science had an important influence on seventeenth-century science but rather why this influence has not been recognized generally. That question remains open for future discussion.

In showing the importance of medieval physics we are not denying the greatness of Galileo's achievement but rather affirming it. It is Galileo's genius which combined these disparate threads in medieval science along with his own contributions into a coherent mechanics. The Argentinian writer Jorge Luis Borges has said, "The fact is that every writer *creates* his own precursors."[168] This is nowhere better demonstrated than in the history we have discussed. It is only because Galileo and Newton created what we call classical mechanics that we can go back and find their medieval precursors among the large number of medieval views on science.[169] It comes as no surprise that these precursors exist.

EPILOGUE

Teachers have always illustrated Sir Isaac Newton's magnanimity to his predecessors by his statement in a letter to Robert Hooke, "If I have seen farther, it is by standing on the shoulders of giants."[170] He is here referring only to Hooke and Descartes but he acknowledges elsewhere the important influence on his work of Galileo, Kepler, and Copernicus. He is definitely not giving any credit to any of his medieval precursors. It is ironic that the expression itself has a long history dating from the Middle Ages.[171] Perhaps that is a fitting way to end this paper.

ACKNOWLEDGMENTS

I would like to thank Professor Edward Grant of the University of Indiana for both his aid and encouragement. He graciously provided large amounts of material which were invaluable in providing both the inspiration for this work and some of its content. I am also indebted to Professor Winifred Wisan, my colleague at the New School of Liberal Arts, and to Dr. Thomas Lyons of the University of Colorado for many helpful discussions and invariably constructive criticism. Much of this work was done while I was on leave at the New School of Liberal Arts of Brooklyn College. I would like to thank Dean Alden Sayres and his colleagues for their hospitality.

I would like to acknowledge the fact that very little of what I have written here is original. It is based on the work of the many writers whom I hope I have acknowledged sufficiently in the notes to this paper. My major contribution has been to organize this material and to collect it in one place.

NOTES

1.   Most physics textbooks do not mention medieval physics at all. Those that do treat it rather briefly and inadequately. Even as historically oriented a text as Eric M. Rogers, *Physics for the Inquiring Mind* (Princeton University Press, Princeton, N.J. 1960) has only a brief mention of Roger Bacon as its total reference to the Middle Ages. John M. Bailey, *Liberal Arts Physics* (W. H. Freeman, San Francisco, 1974) devotes several pages to medieval physics but erroneously attributes the origin of impetus theory to William of Ockham who opposed this theory vehemently. Sometimes the references are quite hostile. Peter J. Brancazio, *The Nature of Physics* (Macmillan, New York, 1975) refers to this period as follows: "During this period science did not just stand still; it fell into total collapse." George Gamow, *Biography of Physics* (Harper & Row, New York, 1961) calls medieval physics "primitive Lysenkoism." Some historians of physics have not treated the medieval period any better. In discussing Archimedes and Galileo, Lagrange wrote, "The interval separating these two great geniuses disappears in the history of mechanics." Quoted in C. Truesdell, *Essays in the History of Mechanics* (Springer-Verlag, New York, 1968), p. 27. Ernst Mach wrote that "dynamics was founded by Galileo" and in discussing free fall motion said, "no part of the knowledge and ideas on this subject with which we are now so familiar, existed in Galileo's time, but . . . Galileo had to create these ideas and means for us." Ernst Mach, *The Science of Mechanics*, translated by T. J. McCormack (Chicago, 1902), pp. 128, 133. Not all historians accept this point of view. Truesdell certainly disagrees with it and Pierre Duhem writes, "When we watch the science of Galileo triumph over the stubborn Peripateticism of a Cremonini, we, being badly informed of

the history of human thought, believe that we are witnessing the victory of a young modern science over medieval philosophy ... in reality, we are witnessing the triumph, long in preparation, of the science which was born at Paris in the fourteenth century, over the doctrines of Aristotle and Averroes which had been restored to honor by the Italian Renaissance." Pierre Duhem, *Etudes sur Léonard de Vinci*, 3e Serie, Les Precurseurs Parisiens de Galilee (Paris, 1913), pp. v-vi. As we shall see, even Duhem did not give sufficient credit to the fourteenth-century work done at Merton College, Oxford or to earlier writers. For other references on medieval physics see reference 4.

2. Thomas S. Kuhn, *The Structure of Scientific Revolutions* (University of Chicago Press, Chicago, 1970), p. 137, pp. 166-67.

3. For an interesting view on the continuity of physics in the medieval period as well as a brief summary of some of the contributions of medieval science see R. A. Uritam, "Medieval Science, the Copernican revolution, and physics teaching," *American Journal of Physics*, Vol. 42, Number 10, October 1974, and the references in his Footnote 9.

4. The study of medieval science is essentially a twentieth-century phenomenon. Pioneering work was done by Pierre Duhem, *Etudes sur Léonard de Vinci* (Paris 1906-13) 3 vols. Duhem's work has been supplemented by Anneliese Maier, *Studien zur Naturphilosophie der Spatscholastik*: I. Die Vorlaufer Galileis in 14, Jahrhundert (Rome, 1949); II. Zwei Grundprobleme der scholastischen Naturphilosophie, (Rome, 1951); III. An der Grenze von Scholastik und Naturwissenschaft, (Rome, 1952); IV. Metaphysische Hintergrunde der spatscholastischen Naturphilosophie, (Rome, 1955); V. Zwischen Philosophie und Mechanik, (Rome, 1958). There are several indispensable reference

works in English: Marshall Clagett, *The Science of Mechanics in the Middle Ages* (University of Wisconsin Press, Madison, 1959) hereafter referred to as Clagett, and Edward Grant (Editor), *A Source Book in Medieval Science* (Harvard University Press, Cambridge, 1974), hereafter Grant, *Source Book*. The instructor who wishes to teach medieval science will have some difficulty. The two best textbooks known to me, A. C. Crombie, *Medieval and Early Modern Science*, 2 vols. (Harvard University Press, Cambridge, 1967) and E. J. Dijksterhuis, *The Mechanization of the World Picture* (Oxford University Press, Oxford, 1961) are both out of print. There is, however, an excellent introductory essay by Edward Grant, *Physical Science in the Middle Ages* (Wiley, New York, 1971). Richard C. Dales, *The Scientific Achievement of the Middle Ages* (University of Pennsylvania Press, Philadelphia, 1973) contains not only excellent introductory material but also translated selections of medieval writers.

5. Isaac Newton, *Principia* (University of California Press, Berkeley, 1960), Motte's translation revised by Cajori, p. 13.

6. Much of medieval science consists of commentaries and critiques of Aristotelian physics. Many of these works can stand on their own as important contributions.

7. Aristotle's concept of motion includes not only local motion which is a change of place but any change from potential to actual being. This includes changes in shape, in size, or in the properties of a body.

8. Thus the ultimate equilibrium in the Aristotelian universe would be a spherical earth surrounded by concentric spherical shells of water, air, and fire surrounded by the heavens.

9. Aristotle, *Physics*, translated by R. P. Hardie and R. K.

Gaye, quoted in Morris R. Cohen and I. E. Drabkin, *A Source Book in Greek Science* (Harvard University Press, Cambridge, 1948), p. 203. Hereafter Cohen and Drabkin.

10. Aristotle, *Physics*, Cohen and Drabkin, pp. 203-4.

11. Edward Grant, "Aristotle's Restriction on His Law of Motion: Its Fate in the Middle Ages," in *L'Aventure de la Science*, Melanges Alexandre Koyré. (Histoire de la Pensée, Vol. XII; Paris, Hermann, 1964), pp. 173-97.

12. Aristotle, *Physics*, Cohen and Drabkin, p. 208.

13. The theory of "antiperistasis" is not due to Aristotle but was supported by Plato or his followers.

14. Aristotle, *Physics*, Cohen and Drabkin, pp. 204-5.

15. Aristotle, *Physics*, Cohen and Drabkin, p. 206.

16. Aristotle, *Physics*, Cohen and Drabkin, p. 206.

17. Aristotle, *Physics*, Cohen and Drabkin, p. 205.

18. "What does it mean to be a precursor or a predecessor? . . . In European history the problem has assumed acute form since the School of Duhem acclaimed Nicholas d'Oresme and other medieval scholars as the precursors of Copernicus, Bruno, Francis Bacon, Galileo, Fermat, and Hegel. Here the difficulty is that every mind is necessarily the denizen of the organic intellectual medium of its own time, and propositions which may look very much alike cannot have had quite the same meaning when considered by minds at very different periods. Discoveries and inventions are no doubt organically connected with the milieu in which they arose. Similarities may be fortuitous. Yet to affirm the true originality of Galileo and his contemporaries

is not necessarily to deny the existence of precursors, so long as that term is not taken to mean absolute priority or anticipation," Joseph Needham, *Science and Civilization in China* (Cambridge University Press, Cambridge, 1962), p. xxvii, and also Uritam, ref. 3. Needham's view is quite reasonable, although we shall see that some of the discoveries we usually attribute to Galileo were in fact done by some of his predecessors.

19. This section is based in part on two papers. Ernest A. Moody, "Galileo and Avempace: The Dynamics of the Leaning Tower Experiment," *Journal of the History of Ideas*, Vol. 12 (1951), pp. 163-93 and 375-422, (hereafter Moody) and Edward Grant, "Motion in the Void and the Principle of Inertia in the Middle Ages," *Isis*, Vol. 55 (1964), pp. 265-92, (hereafter Grant, *Motion in the Void*).

20. Max Jammer, *Concepts of Space* (Harvard University Press, Cambridge, 1969), p. 10.

21. Lucretius, *On the Nature of Things*, translated by Cyril Bailey, Cohen and Drabkin, p. 215.

22. Simplicius, *Commentary on Aristotle's De Caelo*, Cohen and Drabkin, p. 209.

23. This argument that motion in a void is successive because the object cannot be at two places simultaneously appears extensively in medieval discussions of the problem and is known there as the *distantia terminorum*.

24. John Philoponus, *Commentary on Aristotle's Physics*, Cohen and Drabkin, pp. 218-19.

25. It is quite surprising then that Philoponus believes that in air objects fall at the same rate. "For if you let fall from the same height two weights of which one is many times as

heavy as the other, you will see that the ratio of the times required for the motion does not depend on the ratio of the weights, but that the difference in time is a very small one." Philoponus, *Commentary on Aristotle's Physics*, Cohen and Drabkin, p. 220. We see here the same statement as Galileo's mythical Leaning Tower of Pisa experiment. It is unlikely that Philoponus originated this experiment but, as we shall see, Galileo was aware of Philoponus' work and was no doubt influenced by him.

26. John Philoponus, *Commentary on Aristotle's Physics*, Cohen and Drabkin, p. 217.

27. John Philoponus, *Commentary on Aristotle's Physics*, Cohen and Drabkin, pp. 221-22.

28. John Philoponus, *Commentary on Aristotle's Physics*, Cohen and Drabkin, p. 223.

29. S. Pines, "Un précurseur Bagdadien de la Théorie de l'impetus," *Isis*, Vol. 44 (1953), pp. 247-51, quoted in Clagett, p. 510, note 9.

30. Avicenna, *Book of the Healing of the Soul*, translated by Marshall Clagett, Clagett, pp. 510-12.

31. "The *vis insita*, or innate force of matter, is a power of resisting, by which every body, as much as in it lies, continues in its present state, whether it be of rest, or of moving uniformly in a right line. . . . Upon which account this *vis insita* may, be a most significant name, be called inertia (*vis inertiae*) or force of inactivity." Newton, *Principia*, ref. 5, p. 2.

32. Avicenna, Clagett, p. 513.

33. For a more thorough discussion of this problem see Grant, *Motion in the Void*.

34. This definition of *mail* is quite similar to the definition of "impetus" given by John Buridan which is discussed in the next section.

35. This discussion of the acceleration of falling bodies is similar to that given above by Hipparchus. Hipparchus, however, views the force of gravity as a constant, whereas Abu'l-Barakat thinks of the gravity as supplying successive increments of *mail*. A similar explanation appears in the work of Buridan (see the next chapter) althoughBuridan's impetus is regarded as permanent.

36. Abu'l-Barakat, Clagett, p. 547.

37. Moody, p. 414.

38. Clagett, p. 259.

39. Grant, *Source Book*, p. 41.

40. Crombie, ref. 4, p. 60.

41. Clagett, pp. 515-17.

42. Averroes believed these views to be original with Avempace. "Avempace therefore raised doubts in this passage, in two places. . . . And nobody before him had arrived at these questions: and hence he was more profound than any other." Averroes, *The Works of Aristotle with Commentaries by Averroes*, quoted in Moody, p. 192.

43. Averroes, quoted in Grant, *Source Book*, pp. 256-57.

44. Grant has pointed out that Avempace is really stating that $v = v' - r$ where v is the velocity in the medium, v' is the velocity in a void, and r is the loss of velocity due to the resistance of the medium. Grant, *Source Book*, p. 257,

note 17. Nevertheless this is similar enough to Philoponus' law $V = F - R$ for our purposes.

45. Alexander Koyré, *Etudes Galiléennes* (Paris, 1939), pp. 54-60.

46. Moody, pp. 377-80.

47. Moody, p. 378.

48. St. Thomas Aquinas, *Commentary on the Eight Books of the Physics of Aristotle*, Grant, *Source Book*, p. 334.

49. Grant, *Motion in the Void*, p. 271.

50. The Condemnation of 1277, Grant, *Source Book*, p. 48.

51. Peter John Olivi, *Quaestones in secundum librium Sententiarum*, Moody, pp. 385-86.

52. Peter John Olivi, Moody, p. 387.

53. Moody, pp. 387-88.

54. See Crombie, ref. 4, p. 60, for a discussion of the origins of this work.

55. Franciscus de Marchia, *On the Sentences of Peter Lombard*, Moody, p. 392. *The Sentences* was a standard theological text at the time.

56. Franciscus de Marchia, Clagett, p. 527.

57. Franciscus de Marchia, Clagett, pp. 528-29.

58. Franciscus de Marchia, Clagett, p. 530.

59. Nicholas Bonetus, Grant, *Motion in the Void*, p. 274.

60. Galileo, *De Motu*, (*On Motion*), translated by I. E. Drabkin, (University of Wisconsin Press, Madison, 1960), p. 49. Galileo wrote two early works with the same name, a treatise and a dialogue. Professor Drabkin has translated the treatise. The treatment is similar in both works. We shall refer to this work hereafter as Galileo, *De Motu*.

61. Galileo occasionally acknowledges the contributions of his predecessors but is not overly generous. More often than not he mentions them only to refute their arguments. He also has a rather high opinion of his own worth. "You cannot help it, Signor Sarsi, that it was granted to me alone to discover all the new phenomena in the sky and nothing to anybody else." Galileo, The Assayer, quoted in Arthur Koestler, *The Sleepwalkers* (Grosset and Dunlap, New York, 1963), p. 468. Galileo here grants no credit to his contemporaries, much less to his predecessors.

62. Moody, *Galileo and Avempace*, pp. 410-19.

63. Galileo, *De Motu*, see ref. 60.

64. Grant, "Aristotle, Philoponus, Avempace and Galileo's Pisan Dynamics," *Centaurus*, Vol. 11 (1965), pp. 79-95, discusses this point in detail.

65. Galileo, *De Motu*, p. 34.

66. Galileo, *De Motu*, p. 36.

67. Galileo, *De Motu*, pp. 43-44.

68. Galileo, *De Motu*, pp. 78-79. This explanation of projectile motion is quite similar to that of Hipparchus quoted above.

69. For a detailed discussion of some of the ideas in this section, particularly those of Ockham and Buridan, see James F. O'Brien, "Some Medieval Anticipations of Inertia," *The New Scholasticism*, Vol. XLIV, Number 3, Summer 1970, pp. 345-71.

70. William of Ockham, *Expositio super libros Physicorum*, Moody, pp. 397-99.

71. William of Ockham, *Summa Totius Logicae*, Crombie, ref. 4, p. 62.

72. William of Ockham, *Tractatus de Successivis*, Crombie, ref. 4, p. 64.

73. William of Ockham, *Commentary on The Sentences*, Crombie, ref. 4, p. 63.

74. Crombie, ref. 4, p. 65.

75. Marsilius of Inghen, *Questions on the Eight Books of the Physics* (of Aristotle), Clagett, p. 623.

76. Thomas Bradwardine, *De causa Dei contra Pelagium*, quoted in Edward Grant, "Medieval and Seventeenth Century Conceptions of an Infinite Void Space Beyond the Cosmos," *Isis*, Vol. 60, (1969), pp. 39-60.

77. Nicholas Oresme, *Questions on De caelo* (of Aristotle), Latin edition, Grant, ref. 78, p. 48.

78. For a detailed discussion of these influences see Grant, ref. 76.

79. Ockham's views also influenced the kinematic treatment of motion given by Thomas Bradwardine (a contemporary of Ockham at Oxford) and his followers at Merton College,

Oxford. Bradwardine proposed $V = k \log F/R$ (this is, of course, a modern formulation of Bradwardine's Law) as a replacement for the laws of motion of both Aristotle and Avempace. (Thomas Bradwardine: His *Tractatus de Proportionibus*, edited and translated by H. Lamar Crosby, Jr. [University of Wisconsin Press, Madison, 1955]). See also Grant, *Source Book*, pp. 292-305. His followers, John Dumbleton (Fl. 1331-1349), William Heytesbury (1313-1372), and Richard Swineshead (Fl. 1330), all did important work on kinematics. Most importantly they proved the Merton College Mean Speed Theorem which shows that the distance travelled in a time t by a uniformly accelerated body is equal to the distance travelled in that same time t by an object travelling uniformly at the mean speed. (See Clagett, Chapters 4 and 5.) The geometrical proof of this theorem given by Oresme (Clagett, Chapter 6, and Uritam, ref. 3.) is virtually identical to that given by Galileo. Oresme, also, viewed the velocity of a falling body as proportional to time. In his "Questions on the Elements of Euclid," Oresme includes the fact that for a uniformly difform quality (uniformly accelerated motion) if we divide the line segments (time) into equal parts the partial qualities (distances) are as a series of odd numbers (Clagett, p. 344). Oresme does not explicitly apply this statement to motion as I have indicated in the parentheses above. It is clear that this is implied since he regards motion as a special case of his generalized treatment of qualities. He does not, however, explicitly apply this analysis to falling bodies, one of the puzzling problems in the history of science. Several Merton College writers had already noted that the distance travelled in the second half of a time period by a uniformly accelerating body is three times the distance travelled in the first half. (See Clagett, Chapter 5.) The full extent of these studies and their influence on the history of mechanics, particularly on Galileo, is beyond the scope of this paper. It does indicate, however, another important medieval influence on Galileo's work. For an excellent and balanced presentation see Ernest A. Moody, "Galileo and His Precursors," in Carlo L. Golino (Editor), *Galileo Reappraised*

(University of California Press, Berkeley and Los Angeles, 1966), pp. 23-43).

80. John Buridan, *Questions on the Eight Books of the Physics of Aristotle*, Clagett, p. 534. Buridan's attitude toward science and toward Aristotle is indicated at the end of a similar discussion in one of his other works where he concludes, "If you find some way of saving the opinion of Aristotle and the appearances at the same time, I shall gladly adopt that view." Quoted in Dales, ref. 4, p. 121.

81. John Buridan, Clagett, p. 537.

82. John Buridan, Clagett, p. 535.

83. John Buridan, *Questions on the Four Books on the Heavens and the World of Aristotle*, Clagett, p. 561.

84. John Buridan, Clagett, p. 596.

85. We are here neglecting the fact that the earth's rotation is an accelerated motion, and that the arrow would share only the tangential velocity.

86. See Grant, *Motion in the Void*.

87. John Buridan, Grant, *Motion in the Void*, p. 279.

88. John Buridan, Clagett, pp. 560-61.

89. John Buridan, Clagett, p. 560.

90. John Buridan, Clagett, p. 559.

91. Albert of Saxony, *Questions on the [Four] Books on the Heavens and the World of Aristotle*, Clagett, p. 566.

92. Albert of Saxony, Clagett, p. 566.

93. I. B. Cohen, "Galileo's Rejection of the Possibility of Velocity Changing Uniformly with Respect to Distance," *Isis*, Vol. XXVII, pt. 3, no. 149, pp. 231-35, has suggested that Galileo originally derived his knowledge of both fourteenth-century kinematics and dynamics from Albert of Saxony, who assumes the velocity of fall is proportional to distance. Thus he may have interpreted the Merton College Theorem as a space integral rather than a time integral, and may have honestly believed he was the first to demonstrate that the distance of free fall is proportional to the square of the time, even though this had been done by others.

94. Nicholas Oresme, *Commentary on Aristotle's De caelo*, Clagett, p. 554.

95. This view of impetus allows Oresme to explain the false medieval view that a projectile continues to accelerate after it has left the projector. Since the impetus does not immediately decrease to zero, the acceleration also continues.

96. Nicholas Oresme, *On the Book of the Heavens and the World of Aristotle*, Clagett, p. 570.

97. Galileo, *Dialogue Concerning the Two Chief World Systems*, translated by Stillman Drake, (University of California Press, Berkeley and Los Angeles, 1970), p. 22. Hereafter Galileo, *Two World Systems*.

98. Galileo, *Two World Systems*, p. 227. This statement occurs just before another discussion of an object dropped through a hole in the earth.

99. For a thorough discussion of this see Clagett, Chapter 11.

100. Blasius is unusual among science instructors since he was fired from the University of Padua in 1411 "on the grounds that he seemed less fit to teach and that his classes were not attended." Lynn Thorndike, *History of Magic and Experimental Science* (Columbia University Press, New York, 1934), Vol. IV, p. 70.

101. Blasius of Parma, *Questions on the Physics*, Grant, *Motion in the Void*, pp. 289-90, Note 83.

102. Nicholas of Cusa, *Dialogorum de Ludo Globi*, quoted in Max Jammer, *Concepts of Force* (Harvard University Press, Cambridge, 1957), p. 71.

103. The influence of Cusa and other later writers on Galileo's inertial views is discussed by Wohlwill, "Die Entdeckung des Beharrungsgesetzes," *Zeitschrift fur Volkerpsychologie und Sprachwissenschaft*, Vol. 14, pp. 365-410 (1883) and Vol. 15, pp. 70-135, 337-87, (1884). This discussion was written before the work of the medieval scientists we have been discussing was available.

104. Leonardo da Vinci, Manuscripts, Clagett, pp. 572-73.

105. Domingo de Soto, *Questions on the Eight Books of Aristotle's Physics*, Clagett, pp. 555-56.

106. Benedetti, *Book of Various Mathematical and Physical Ideas*, quoted in *Mechanics in Sixteenth-Century Italy*, translated by Stillman Drake and I. E. Drabkin (University of Wisconsin Press, Madison, 1969), pp. 186-87.

107. Benedetti, ref. 106, p. 189.

108. Galileo, *Dialogues Concerning Two New Sciences*, translated by Crew and de Salvio (Dover, New York, 1954), p. 254. Hereafter *Two New Sciences*.

109. Galileo, *Two New Sciences*, pp. 250-51.

110. Galileo, *Two New Sciences*, p. 181.

111. Galileo, *Two World Systems*, pp. 193-94.

112. Galileo, *Two World Systems*, p. 148.

113. See Wohlwill, ref. 103.

114. Clagett, pp. 414, 651.

115. Clagett, p. 651.

116. Moody, p. 166.

117. As we shall see, Descartes regarded motion as relative and, in fact, defined motion in a way similar to Ockham. Newton believed in absolute motion, but in his principle of inertia treats rest and uniform motion as equivalent states which is a view closely related to the relativity of motion.

118. Simplicius, *Commentary on Aristotle's De caelo*, Cohen and Drabkin, pp. 106-7.

119. Avempace, quoted in Averroes, Grant, *Source Book*, p. 257.

120. Franciscus de Marchia, Clagett, p. 530.

121. Buridan, Clagett, p. 561.

122. Anonymous, *On the Elements*, translated by R. C. Dales, quoted in Dales, ref. 4, p. 57.

123. Buridan, *Questions on the Four Books of the Heavens and the World of Aristotle*, Clagett, pp. 594-95.

124. We see here the influence of another medieval scientist, Witelo, on Oresme's views. Grant points out that Witelo took his view from the Optics of Alhazen, Grant, *Source Book*, p. 505, Note 39.

125. Oresme, *Le Livre du ciel et du monde*, translated by Albert D. Menut, Grant, *Source Book*, pp. 504-5.

126. Nicolaus Copernicus, *De Revolutionibus orbium coelestium*, translated by Charles Wallis, Grant, *Source Book*, p. 511.

127. Galileo, *Two World Systems*, p. 116.

128. Oresme, Grant, *Source Book*, p. 505.

129. Copernicus, Grant, *Source Book*, p. 514.

130. Galileo, *Two World Systems*, pp. 186-87.

131. Buridan, Clagett, p. 596.

132. Oresme, Grant, *Source Book*, p. 505.

133. Copernicus, Grant, *Source Book*, p. 514.

134. Galileo, *Two World Systems*, p. 142.

135. Buridan, Clagett, p. 596.

136. Oresme, Grant, *Source Book*, p. 505.

137. Galileo, *Two World Systems*, p. 174.

138. It is interesting that Oresme chooses an arrow for his projectile while Galileo uses a cannon ball. We see the effect of the advances in military technology over the two hundred fifty years separating the two authors. We also notice that

the projectile question does not seem to be discussed by Copernicus who is primarily an astronomer.

139. Oresme, Grant, *Source Book*, p. 510.

140. Grant, *Physical Science in the Middle Ages*, ref. 4, pp. 68-69.

141. Clagett, pp. 588-89, 637-38, and Chapter 11.

142. For a detailed study of this medieval influence on Galileo's work see W. L. Wisan, "The New Science of Motion: A Study of Galileo's De motu locali," *Archive for the History of Exact Sciences*, Vol. 13, Number 2/3, (1974), pp. 103-6. Medieval statics is studied extensively in Ernest A. Moody and Marshall Clagett, *The Medieval Science of Weights* (University of Wisconsin Press, Madison, 1952). See also Clagett, ref. 4, Chapters 1 and 2.

143. Hero of Alexandria, *Mechanics*, quoted in A. G. Drachmann, *The Mechanical Technology of Greek and Roman Antiquity* (Hafner, London, 1963), p. 46.

144. The treatise on statics and machines, "Mechanica" was originally attributed to Aristotle, but seems to have been written by one of his followers.

145. Jordanus of Nemore, *De ratione ponderis*, Grant, *Source Book*, pp. 212-13.

146. Blasius of Parma and Leonardo da Vinci also consider motion on an inclined plane but their discussion is more closely related to the brachistochrone problem, or the path of shortest descent.

147. W. L. Wisan, ref. 142, p. 149.

148. Benedetti, quoted in Drake and Drabkin, ref. 106, p. 184.

149. Benedetti, quoted in Drake and Drabkin, ref. 106, p. 185.

150. Galileo, *De Motu*, p. 64.

151. Galileo, *De Motu*, pp. 65-66.

152. Galileo's reference to a plane parallel to the horizon again implies his view that a horizontal plane is actually the spherical surface of the earth.

153. Galileo, *On Mechanics*, translated by Stillman Drake (University of Wisconsin Press, Madison, 1960), p. 171.

154. Galileo, *Letters on Sunspots*, quoted in Stillman Drake, trans. *Discoveries and Opinions of Galileo* (Doubleday Anchor Books, Garden City, New York, 1957).

155. Galileo, *Two New Sciences*, p. 215.

156. One of Galileo's applications of motion on an inclined plane and his inertial views, the "double distance" rule, also shows the influence of medieval thought. Galileo states this rule as follows, "Hence we can infer that, if, after descent along the inclined plane AC of the adjoining figure, the motion is continued along a horizontal line, such as CT, the distance traversed by a body, during a time equal to the time of fall through AC, will be exactly twice the distance AC." (Galileo, *Two New Sciences*, p. 214). He is clearly assuming that motion along a horizontal line (parallel to the earth's surface) is constant. Galileo's first discussion of this rule occurs after he has discussed a "double velocity" rule. This rule states that if a body is accelerated from rest from point A to point B and then travels the distance AB again with a uniform speed equal to the maximum speed attained in the accelerated motion, the distance AB will be travelled

twice as quickly. This seems to be an application of the Merton College Mean Speed Theorem and Galileo's geometrical proof of this rule looks quite similar to the geometrical method of Oresme and Casali. He then derives the double distance rule. (See W. L. Wisan, ref. 142, pp. 204-6). A second reference occurs in his *Dialogue on the Two World Systems*. He considers the motion of an object dropped into a hole drilled through the center of the earth. He assumes, incorrectly, that the speed will increase uniformly from 0 to 10 and then back to 0 as it reaches the other side. This example, in fact, follows the statement on this type of motion which we cited earlier as being similar to that of Oresme and Albert of Saxony. He then proceeds to prove the "double distance" rule using geometrical methods quite similar to those of medieval scientists. Galileo even associates the distance travelled with the "total sum of speeds," the area under the velocity-time graph, which is quite similar to the fourteenth-century usage of associating distance with *velocitatis totalis* or total velocity (Galileo, *Two World Systems*, pp. 226-30). A similar geometrical proof follows his statement of the "double distance" rule in the *Dialogue Concerning Two New Sciences*, quoted above.

157. In 1641, the Count d'Allais placed a galley at the disposal of Gassendi, who performed the experiment in the harbor of Marseille. Alexandre Koyré, *Newtonian Studies* (Harvard University Press, Cambridge, 1965), p. 176. Hereafter Koyré.

158. Gassendi, *De motu impresso a motore translato* (Paris 1641), quoted in Koyré, pp. 178-79.

159. Edict of Paris, Grant, *Source Book*, p. 48.

160. Gassendi, *De motu*, Koyré, pp. 186-87.

161. Descartes, *Le Monde*, Koyré, p. 71-72.

162. Descartes, *Principles of Philosophy*, Koyré, p. 80.

163. Descartes, *Principles of Philosophy*, Koyré, p. 81.

164. Descartes, *Principles of Philosophy*, Koyré, p. 75.

165. Descartes, *Principles of Philosophy*, Koyré, p. 75.

166. Descartes, *Le Monde*, Koyré, p. 73.

167. Isaac Newton, *Principia*, ref. 5, p. 13.

168. Jorge Luis Borges, "Kafka and His Precursors," *Labyrinths*, edited by Donald A. Yates and James E. Irby (New Directions, New York, 1964), p. 201.

169. For a virtually identical view see Ernest A. Moody, "Galileo and His Precursors," ref. 79.

170. Isaac Newton, *Letter to Robert Hooke*, Koyré, p. 227.

171. The earliest reference I have found to this idea appears in the statement of the first century B.C. general, Didacus Stella, who says, "Pigmies placed on the shoulders of giants see more than the giants themselves." Quoted in Lucan (A.D. 39-65), *The Civil War*, II,10. The same view is expressed by Bernard of Chartres, Chancellor of Chartres from about 1119 to 1126, who in comparing modern scholars to the ancients said they were as dwarfs standing on the shoulders of giants. A similar, although not equivalent, view is shown in four of the stained glass windows in the Cathedral of Chartres. The dependence of the New Testament on the Old is illustrated by the four authors of the Gospels, Matthew, Mark, Luke, and John shown sitting on the shoulders of four Old Testament prophets, Isaiah,

Daniel, Jeremiah, and Ezekial. Obviously this was not intended to show the evangelists as dwarfs in comparison to the prophets and this is indicated by the similar size of the figures. This view is probably closer to that of Newton, who certainly regarded his own achievements as at least the equal of those of Descartes and Hooke. Raymond Klibansky, "Standing on the Shoulders of Giants," *Isis*, 26, (1936), pp. 147-49, discusses the medieval history of this expression. An extensive and witty history of this expression is given in Robert K. Merton, *On the Shoulders of Giants* (The Free Press, New York, 1965). Merton points out that the Didacus Stella enshrined in Bartlett's "Familiar Quotations" is not the first century B.C. general, but is rather Diego de Estella, a sixteenth-century writer. He attributes the origin of the aphorism to Bernard of Chartres who, of course, stands on the shoulders of those who came before him, notably Priscian.